AUSTRALIAN MATHEMATICAL SOCIETY LECTURE SERIES

Editor-in-Chief: Dr. S.A. Morris, Department of Mathematics, La Trobe University,
Bundoora, Victoria 3083, Australia

Subject Editors:
Professor C.J. Thompson, Department of Mathematics, University of Melbourne,
Parkville, Victoria 3052, Australia
Professor C.C. Heyde, Department of Statistics, University of Melbourne,
Parkville, Victoria 3052, Australia
Professor J.H. Loxton, Department of Pure Mathematics, University of New South Wales,
Kensington, New South Wales 2033, Australia

for Zeny
wife, mother and
γιαγια

QA
611.28
G55
1987

Australian Mathematical Society Lecture Series. 3

Introduction to the Analysis of Metric Spaces

J.R. Giles
Faculty of Mathematics,
The University of Newcastle, New South Wales

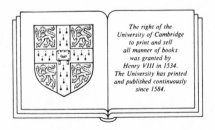

*The right of the
University of Cambridge
to print and sell
all manner of books
was granted by
Henry VIII in 1534.
The University has printed
and published continuously
since 1584.*

CAMBRIDGE UNIVERSITY PRESS

Cambridge

New York New Rochelle Melbourne Sydney

Published by the Press Syndicate of the University of Cambridge
The Pitt Building, Trumpington Street, Cambridge CB2 1RP
32 East 57th Street, New York, NY 10022, USA
10, Stamford Road, Oakleigh, Melbourne 3166, Australia

First Published 1987

Printed in Great Britain at the University Press, Cambridge

Library of Congress cataloging in publication data applied for

British Library cataloguing in publication data

Giles, John R.
Introduction to the analysis of metric spaces.---(Australian Mathematical
Society lecture series; 3)
1. Metric spaces
I. Title   II . Series
515.7'3   QA611.28

ISBN   0 521 35051 4   hard covers
ISBN   0 521 35928 7   paperback

CONTENTS

Open sets and their properties, base for a
topology;   equivalent metrics;   relation to
closed sets;   the interior of a set;   the
characterisation of continuous mappings by
inverse images;   topological compactness;
separability, the normal topological structure;
Exercises.

PREFACE

This text is designed as a basic introductory course in the
analysis of metric and normed linear spaces for undergraduate students.
It is aimed at providing the abstract analysis components for the degree
course of a student majoring in mathematics or an honours student majoring
in science or engineering.

It is assumed that such students will have completed a first
course in real analysis or a course in calculus which has been carefully
developed with attention given to the real analysis foundations. The text

Calculus by Michael Spivak
Benjamin, 1967

is universally acknowledged as presenting just such a calculus course with
an eye on rigour.

It is also assumed that the student will have some background
in elementary linear algebra.

Such an analysis course as presented here would be included in
the lecture programme for a standard undergraduate course no earlier than
in the second year. The second year undergraduate programme in Australian
universities certainly includes the calculus of functions of many variables,
introductory complex analysis and further linear algebra. So it is
assumed that such study will be progressing at least concurrently with
this course in analysis.

The author follows the educational style set in

Foundations of Modern Analysis by J. Dieudonné
Academic Press, 1960

and which has been the dominant practice in British universities, of
studying the analysis of metric spaces in detail before introducing
general topological spaces. After all, the spaces which a mathematician
is most likely to face are metric spaces and it is preferable from a
teaching point of view to work from the more familiar and concrete to the
less familiar and abstract. This approach also enables us to gain an
early appreciation of the abstract structural nature of much of real
analysis: for example, uniform convergence is easier to grasp when we
visualise it as convergence in a special function space. But worthwhile
applications are also more immediate and this is important in selling
the advantages of abstract analysis.

It is important that students majoring in mathematics gain
some familiarity with the axiomatic method in analysis for it provides a
logically tight investigation of a basically simple abstract structure
which manifests itself in a number of diverse examples. The justification
for studying abstract metric and normed linear spaces is that these are
the fundamental analysis structures underlying many problems in analysis.
To demonstrate this we present in the first chapter a wealth of example
spaces. In subsequent chapters, after introducing a concept in the
general structure we illustrate it in a variety of example spaces.

The mathematician must be able to detect the significant
underlying structure in a problem situation and then be able to apply the
properties of the general structure to the solution of the specific
problem. One of the most spectacular successes in applying metric space
theory is in the study of the contraction mapping principle: in the
second chapter where we introduce convergence in metric spaces, we go on
to demonstrate quite early the great power of the method in applying
Banach's Fixed Point Theorem to a range of quite disparate situations.
In Chapter IV we aim to communicate some appreciation of the beauty of
the existence approach to a proof of the Weierstrass Approximation Theorem
and demonstrate the techniques of generalising the proof to a more
general situation in the Stone-Weierstrass Theorem. By this stage the
student should be impressed with the considerable achievements made by
exploiting the abstract structural nature of a problem.

An important feature of this text is that normed linear spaces are introduced from the beginning as a special subfamily of metric spaces. After all most of the metric spaces we deal with in examples are normed linear spaces or subsets of normed linear spaces and it is worthwhile pointing out this structure at the outset. Moreover there are some simplifications which follow from a recognition of normed linear structure: the triangle inequality is easier to handle, the spheres and balls are all translations of scaled images of the unit sphere at the origin and the norm topology accords more with our intuition. Many topics introduced for metric spaces have particular implications for normed linear spaces: for example, with closure we discuss the closure of linear subspaces, with continuity it is natural to enquire into the continuity of linear mappings, and compactness is applied to characterising finite dimensionality.

The basic technique we rely on is the sequential method: for example, our basic definition of cluster points is by convergent sequences, we also point to the importance of the sequential characterisation of continuity of mappings and we derive the properties and applications of compactness from sequential compactness. Furthermore, any explicit reference to the metric topology is deferred to the end of the book. The author has found from teaching courses in the analysis of metric spaces that the early introduction of the notion of open sets and their properties is often an abstraction which side tracks the main thrust of the course without noticeably facilitating the development of the theory. Where a topological account is important, as in a fuller discussion of compactness, it is quite satisfactory to work with the open ball base for the topology. The concluding chapter concerns the metric topology and reviews the earlier material putting it in a topological setting to fit into a further course on the analysis of general topological spaces.

From the earliest development of analysis for normed linear spaces it is natural to enquire about finite dimensional spaces and especially so because the theory is a generalisation from that more familiar situation. It has always seemed rather unnatural to defer a discussion of finite dimensional spaces in general until compactness arguments have been developed. In this text, on the other hand, as each new concept is introduced we determine the particular form it takes for general finite dimensional normed linear spaces as a significant example

situation. Using the local compactness of the real number system,
we establish that in finite dimensional normed linear spaces, convergence
is equivalent to co-ordinatewise convergence and we use this as the
fundamental property from which the finite dimensional analysis properties
are derived. It is important for the student to understand early that
although the study of normed linear spaces springs from finite dimensional
spaces the interest in the finite dimensional situation lies essentially
in their linear algebra and their being isomorphic to Euclidean or
Unitary spaces.

Some topics are notable absent from this course. Connectedness
is not mentioned except that we do use the connectedness property of
continuous real functions from real analysis. Otherwise it is better
dealt with from a topological point of view and is not in general
applicable to the course we are considering. We make no reference to
example spaces using the Lebesgue integral because it is the opinion of
the author that a first course in abstract analysis should not be delayed
until the student has studied Lebesgue integration. When integration
theory comes to be studied the student should be in a position to
appreciate the normed linear space structures which are there.

This text is an introduction to what is traditionally called
functional analysis. But here we make no use of the Axiom of Choice in
its most general form. The analysis of normed linear spaces is developed
up to the point where, with the Axiom of Choice in the form of Zorn's
lemma, the Hahn-Banach Theorem can be established. The theory of normed
linear spaces as abstract entities begins from that point. Such a study
examines duality and the theory of linear operators.

We have not introduced abstract Hilbert spaces and their axiom
system reckoning that their special inner product structure finds its
major significance not in its metric space analysis but in an
exploration of duality and operator theory, subjects which properly belong
to a sequel to this text.

In the preface to *Foundations of Modern Analysis*, J. Dieudonné says that he makes "no appeal whatsoever to 'geometrical intuition' at least in the formal proofs ... which we have emphasised by deliberately abstaining from introducing any diagram in the book.". At the time of his writing Dieudonné had a point to make. Now although formal proofs should not rely on diagrams, 'geometrical intuition' does play a vitally important part in creating mathematical ideas and in setting up the rough frame for a proof even in an abstract setting. Moreover, for abstract spaces it is sensible for the student to develop the habit of asking about the form his problem takes, say, in the Euclidean plane, and that often implies a visualisation in terms of Euclidean geometry. To ask a student to work mathematically without using geometrical intuition is like asking him to untie a knot blindfolded. Especially is this so for an abstract analysis course aimed at middle course undergraduates. So this text has a limited number of diagrams which are intended to be geometrically suggestive of the formal material contained in the text.

At the end of each section is a graded selection of exercises which follow generally the order of presentation of the material in the section. A basic difficulty for students beginning a course in abstract analysis is to detect the abstract structural properties in concrete example spaces so to help the student gain facility in doing this, the first exercises in each set are designed to follow a basic concept through several example spaces. Later exercises in each set are conceptually more sophisticated and are intended to introduce a diversity of areas where the general theory is applied.

Although we expect a mathematics text to have a logical development, scarcely ever is such a text read as a novel from beginning to end and certainly it is never reread that way. So a mathematics text without an adequate index is a continual source of frustration to the reader. The index in this text is intentionally detailed giving references to all the significant places where a particular concept is used.

To begin a mathematics text with a preliminary chapter cataloguing prerequisite knowledge is hardly a way to induce initial excitement in the reader. Besides, some or all of such material will be well known by a great many readers. So in this text such material is presented in an appendix.

Most of the material in this lecture course can be covered in between 25-30 lectures given an adequate tutorial programme. However, the material can be readily tailored for various length and content requirements. The preferred use of sequential methods implies that a shorter course using only these methods would be one which concludes in Section 8 before the alternative characterisations of compactness are introduced. Of course the economies of lecturing would demand that not all example spaces be introduced in class and the applications in Section 5 could be curtailed if it is felt that such material overlaps too much with other courses.

However, any course which treats normed linear spaces even in an introductory way as we do here, should discuss linear mappings. A lecturer who proposes to cut Section 7 out of his course should consider the effect of his doing so on the later material in the text.

There are many texts giving more detailed accounts of different sections of the course presented here. Although this text takes its initial inspiration from the approach of J. Dieudonné's book mentioned above, his presentation is generally difficult for a middle year undergraduate student and it covers more ground with more sophistication.

A very useful book written from the same viewpoint but with the development of further analysis of normed linear spaces in mind is

Elements of Functional Analysis by A.L. Brown and A. Page
van Nostrand Reinhold, 1970.

A text on the analysis of metric spaces which is aimed to bring an abstract approach into earlier undergraduate courses is

Real Analysis: an introduction by A.J. White
Addison-Wesley, 1968.

This text is useful in that it puts real analysis in an abstract setting, but there is no explicit reference to normed linear spaces.

Texts such as

Topology and Modern Analysis by George F. Simmons
McGraw-Hill, 1963 and

Topology and Normed Spaces by G.J.O. Jameson
Chapman-Hall, 1974

are both very readable as introductory texts and might be used as
references but they develop from a topological point of view.
    A text which is very popular with engineers is

Introductory Functional Analysis with Applications
by E. Kreysig
John Wiley & Sons, 1978.

His text contains much more material than ours. Kreysig's aim has been
to reduce the topology prerequisites for the course. This will make his
text an accessible reference for our course.

    The author has given lectures on a course such as this to
second year mathematics major students at the University of Newcastle for
a period of 10 years. The lecture notes were produced in duplicated form
in 1974 but subsequent experience has modified the initial approach
considerably to arrive at the form presented here.
    Thanks are due to my colleagues in the department for their
conversations over the years which have had their effect on the final
result, especially Michael Hayes who has always been a steadying
influence directing me to the realities of the teaching situation for the
average students.

    I am indebted to Anne Feletti and to Jan Garnsey who have so
competently created typed order out of my handwritten manuscript. I wish
to express my thanks to the Editors and referees for their encouragement
and assistance in the preparation of this contribution to the Lecture
Series of the Australian Mathematical Society.

J.R. Giles
Associate Professor in Mathematics
The University of Newcastle
N.S.W., Australia

I. METRIC SPACES AND NORMED LINEAR SPACES

After some experience with real and complex analysis it becomes apparent that the development of the theory depends to a great extent simply on the notion of distance between the numbers.

For example, in either number system, when we have a sequence $\{x_n\}$ converging to a limit x we commonly write $|x_n - x| \to 0$ as $n \to \infty$ and say that the distance between x_n and x tends to zero as n tends to infinity. Although the same statement applies in the two different number systems, the meaning of the modulus sign $|\cdot|$ depends on the number system where it is used.

This observation suggests that we generalise the analysis of the real or complex numbers to the analysis of metric spaces. A metric space is a non-empty set where the distance between any two points is specified. The notion of distance has to retain those properties of distance used in the real and complex number systems which are evidently vital for the development of a sensible analysis.

1. DEFINITIONS AND EXAMPLES

1.1 Definition. Given a non-empty set X, a distance function d on X, called a *metric* for X, is a function which assigns to each pair of points a real number, (or formally, $d : X \times X \to \mathbb{R}$), satisfying the following properties:

For all x,y $\in$ X

    (i) $d(x,y) \geq 0$

    (ii) $d(x,y) = 0$ if and only if $x = y$

    (iii) $d(x,y) = d(y,x)$,

and for all x,y,z $\in$ X

    (iv) $d(x,y) \leq d(x,z) + d(y,z)$, (the triangle inequality)

A non-empty set X with a metric d is denoted by (X,d) and is called a *metric space*. Different metrics could be defined on the same set giving rise to different metric spaces.

   With such a general definition we can expect to have some situations where our metric intuition is strained. The following example is important as such an extreme case. As we introduce new concepts this space will be useful in testing our definitions against our intuition.

1.2 Example. For any non-empty set X the *discrete metric* d is defined by

$$d(x,y) = 0 \quad \text{if } x = y$$
$$ = 1 \quad \text{if } x \neq y$$

We may consider this to be the roughest of metrics; given any $x \in X$, it is simply a measure of coincidence with x. □

   Most of the metric spaces we will consider are also linear spaces and it is frequently of advantage to take this into account. In most of these cases, the metric is generated by a simpler function called a norm which assigns a length to each vector in the linear space.

1.3 Definition. Given a linear space X over $\mathbb{R}$ (or $\mathbb{C}$), a *norm* $\|\cdot\|$ for X is a function on X which assigns to each element a real number, (or formally, $\|\cdot\| : X \to \mathbb{R}$), satisfying the following properties:
For all $x \in X$

   (i) $\|x\| \geq 0$
   (ii) $\|x\| = 0$ if and only if $x = 0$
   (iii) $\|\lambda x\| = |\lambda| \|x\|$ for any scalar λ,

and for all $x,y \in X$

   (iv) $\|x+y\| \leq \|x\| + \|y\|$ (the triangle inequality)

A linear space X with a norm $\|\cdot\|$ is denoted by $(X,\|\cdot\|)$ and is called a *normed linear space*. Again, different norms could be defined on the same linear space giving rise to different normed linear spaces.

1.4 Remark. Given a normed linear space $(X, \|\cdot\|)$, it is clear that the function $d : X \times X \to \mathbb{R}$ defined by

$$d(x,y) = \|x-y\|$$

is a metric for X, and we call this *the metric generated by the norm* $\|\cdot\|$. So then every normed linear space is a metric space under the metric generated by its norm. □

1.5 Examples. $(\mathbb{R}, |\cdot|)$ and $(\mathbb{C}, |\cdot|)$

The set of real numbers $\mathbb{R}$ (the set of complex numbers $\mathbb{C}$) is a normed linear space with norm given by the modulus; that is,

$$\|x\| = |x|.$$

We call this the *usual norm* for $\mathbb{R}$ (or $\mathbb{C}$) and it generates the *usual metric*

$$d(x,y) = |x-y|.$$

These are the spaces we are familiar with in real and complex analysis. □

1.6 Examples. $(\mathbb{R}^n, \|\cdot\|_2)$ and $(\mathbb{C}^n, \|\cdot\|_2)$

The set of ordered n-tuples of real numbers $\mathbb{R}^n$ (of complex numbers $\mathbb{C}^n$) is a normed linear space with norm $\|\cdot\|_2$ defined as follows: For $x \equiv (\lambda_1, \lambda_2, \ldots, \lambda_n)$,

$$\|x\|_2 = \sqrt{(|\lambda_1|^2 + |\lambda_2|^2 + \ldots + |\lambda_n|^2)}.$$

We call this the *Euclidean norm* for $\mathbb{R}^n$ (the *Unitary norm* for $\mathbb{C}^n$) and it generates the *Euclidean metric* for $\mathbb{R}^n$ (the *Unitary metric* $\mathbb{C}^n$). We call $(\mathbb{R}^n, \|\cdot\|_2)$ *Euclidean n-space* and $(\mathbb{C}^n, \|\cdot\|_2)$ *Unitary n-space.*

The only norm property which provides any difficulty to verify is the triangle inequality. The proof of this can be derived from the Cauchy-Schwarz inequality.

<u>1.7 Lemma.</u>   <u>The Cauchy-Schwarz inequality</u>

In the linear space $\mathbb{R}^n$ *(or* $\mathbb{C}^n$*), for any* $x \equiv (\lambda_1, \lambda_2, \ldots, \lambda_n)$ *and* $y \equiv (\mu_1, \mu_2, \ldots, \mu_n)$,

$$\sum_{k=1}^{n} |\lambda_k \mu_k| \leq \sqrt{\left(\sum_{k=1}^{n} |\lambda_k|^2\right)} \sqrt{\left(\sum_{k=1}^{n} |\mu_k|^2\right)} .$$

<u>Proof.</u>   For any positive real numbers a and b

$$2ab \leq a^2 + b^2 .$$

Therefore, given non-zero x and y, for each $k \in \{1, 2, \ldots, n\}$, putting

$$a \equiv \frac{|\lambda_k|}{\sqrt{\left(\sum_{k=1}^{n} |\lambda_k|^2\right)}} \quad \text{and} \quad b \equiv \frac{|\mu_k|}{\sqrt{\left(\sum_{k=1}^{n} |\mu_k|^2\right)}}$$

and summing the consequent inequalities we have

$$\frac{\sum_{k=1}^{n} |\lambda_k \mu_k|}{\sqrt{\left(\sum_{k=1}^{n} |\lambda_k|^2\right)} \sqrt{\left(\sum_{k=1}^{n} |\mu_k|^2\right)}} \leq 1. \quad \square$$

<u>Proof of the triangle inequality in 1.6.</u>

$$\begin{aligned}
\|x+y\|_2^2 &= \left(\sum_{k=1}^{n} |\lambda_k + \mu_k|^2\right) \\
&\leq \sum_{k=1}^{n} |\lambda_k|^2 + 2\sum_{k=1}^{n} |\lambda_k \mu_k| + \sum_{k=1}^{n} |\mu_k|^2 \\
&\leq \sum_{k=1}^{n} |\lambda_k|^2 + 2\sqrt{\left(\sum_{k=1}^{n} |\lambda_k|^2\right)} \sqrt{\left(\sum_{k=1}^{n} |\mu_k|^2\right)} + \sum_{k=1}^{n} |\mu_k|^2
\end{aligned}$$

by the Cauchy-Schwarz inequality

$$= (\|x\|_2 + \|y\|_2)^2 . \quad \square$$

The next two sets of examples are formed by taking different norms on the same underlying linear space $\mathbb{R}^n$ (or $\mathbb{C}^n$).

<u>1.8 Examples.</u> $(\mathbb{R}^n, \|\cdot\|_1)$ and $(\mathbb{C}^n, \|\cdot\|_1)$

$\mathbb{R}^n$ (or $\mathbb{C}^n$) is a normed linear space with norm $\|\cdot\|_1$ defined as follows:

For $x \equiv (\lambda_1, \lambda_2, \ldots, \lambda_n)$,

$$\|x\|_1 = |\lambda_1| + |\lambda_2| + \ldots + |\lambda_n|.$$

The triangle inequality follows from the triangle inequality in $(\mathbb{R}, |\cdot|)$:

For any $x \equiv (\lambda_1, \lambda_2, \ldots, \lambda_n)$ and $y \equiv (\mu_1, \mu_2, \ldots, \mu_n)$,

$$\|x+y\|_1 = \sum_{k=1}^{n} |\lambda_k + \mu_k|$$

$$\leq \sum_{k=1}^{n} |\lambda_k| + \sum_{k=1}^{n} |\mu_k|$$

$$= \|x\|_1 + \|y\|_1. \quad \square$$

<u>1.9 Examples.</u> $(\mathbb{R}^n, \|\cdot\|_\infty)$ and $(\mathbb{C}^n, \|\cdot\|_\infty)$

$\mathbb{R}^n$ (or $\mathbb{C}^n$) is a normed linear space with norm $\|\cdot\|_\infty$ defined as follows:

For $x \equiv (\lambda_1, \lambda_2, \ldots, \lambda_n)$,

$$\|x\|_\infty = \max\{|\lambda_k| : k \in \{1, 2, \ldots, n\}\}.$$

We call this the *supremum* (or *uniform*) norm for $\mathbb{R}^n$ (or $\mathbb{C}^n$).

We check that the triangle inequality holds:

For any $x \equiv (\lambda_1, \lambda_2, \ldots, \lambda_n)$ and $y \equiv (\mu_1, \mu_2, \ldots, \mu_n)$, if $\max\{|\lambda_k + \mu_k| : k \in \{1, 2, \ldots, n\}\} = |\lambda_j + \mu_j|$ then

$$\|x+y\|_\infty = |\lambda_j + \mu_j| \leq |\lambda_j| + |\mu_j|$$

$$\leq \max\{|\lambda_k| : k \in \{1, 2, \ldots, n\}\} + \max\{|\mu_k| : k \in \{1, 2, \ldots, n\}\}$$

$$= \|x\|_\infty + \|y\|_\infty. \quad \square$$

We now consider a particular abstract normed linear space where many concrete spaces can be derived as special cases.

<u>1.10 Example.</u> $(B(X), \|\cdot\|_\infty)$

For any non-empty set X we denote by $B(X)$ the set of bounded real (complex) functions on X. $B(X)$ is a real (complex) linear space under pointwise definition of addition and multiplication by a scalar; that is, for any $f, g \in B(X)$,

$$(f+g)(x) \equiv f(x) + g(x) \quad \text{for all } x \in X$$

and for any $f \in B(X)$ and scalar λ

$$(\lambda f)(x) \equiv \lambda f(x) \quad \text{for all } x \in X.$$

$B(X)$ is a normed linear space with norm $\|\cdot\|_\infty$ defined by

$$\|f\|_\infty = \sup\{|f(x)| : x \in X\}.$$

We call this the *supremum* (or *uniform*) norm for $B(X)$. We notice that Example 1.9 is the special case when $X \equiv \{1, 2, \ldots, n\}$. We check that the triangle inequality holds in the general case:
For any $f, g \in B(X)$,

$$|(f+g)(x)| \leq |f(x)| + |g(x)| \quad \text{for every } x \in X$$

so $$\sup\{|(f+g)(x)| : x \in X\}$$

$$\leq \sup\{|f(x)| : x \in X\} + \sup\{|g(x)| : x \in X\}. \quad \square$$

Of particular interest are the following two special cases.

<u>1.11 Example.</u> $(m, \|\cdot\|_\infty)$

When $X = \mathbb{N}$, $B(\mathbb{N})$ is the linear space of bounded sequences usually denoted by m (or sometimes ℓ_∞). In this case the norm $\|\cdot\|_\infty$ takes the following form:
For $x \equiv \{\lambda_1, \lambda_2, \ldots, \lambda_n, \ldots\}$,

$$\|x\|_\infty = \sup\{|\lambda_n| : n \in \mathbb{N}\}. \quad \square$$

1.12 Example. $(B(J), \|\cdot\|_\infty)$ and $(B[a,b], \|\cdot\|_\infty)$

When $X = J$ an interval of real numbers, $B(J)$ is the linear space of bounded functions on J. We will be specially interested in the particular case when $J = [a,b]$ a bounded closed interval. ☐

Many significant metric and normed linear spaces are related to others as subspaces.

1.13 Definitions. Given a metric space (X,d) and Y a non-empty subset of X it is clear that d is also a metric for Y, (that is, the restriction of d to $Y \times Y$ is a metric for Y). We denote this restriction by $d|_Y$ and call it the *relative metric* induced by d on Y. We call $(Y, d|_Y)$ a *metric subspace* of (X,d).
Given a normed linear space $(X, \|\cdot\|)$ and a linear subspace Y of X it is clear that the restriction of the norm $\|\cdot\|$ to Y is also a norm for Y. We denote this restriction by $\|\cdot\|_Y$ and call $(Y, \|\cdot\|_Y)$ a *normed linear subspace* of $(X, \|\cdot\|)$.

Very often a metric space where there is no linear structure, can be considered as a metric subspace of a normed linear space. For example, it is quite natural to consider the interval $[a,b]$ as a metric subspace of $(\mathbb{R}, |\cdot|)$ where the usual metric is restricted to $[a,b]$.

The following examples are significant subspaces of those given in Examples 1.11 and 1.12.

1.14 Example. $(c_0, \|\cdot\|_\infty)$

The set c_0 of sequences which converge to zero is a linear subspace of m, and $(c_0, \|\cdot\|_\infty)$ is a normed linear subspace of $(m, \|\cdot\|_\infty)$.

1.15 Examples. $(C(J), \|\cdot\|_\infty)$ and $(C[a,b], \|\cdot\|_\infty)$

The set $C(J)$ of bounded continuous real functions on an interval J is a linear subspace of $B(J)$. This follows from the algebra of continuous functions in real analysis. In particular, when $J \equiv [a,b]$ a bounded closed interval, since a continuous function on a bounded closed interval is bounded, so $C[a,b]$ is the linear space of continuous real functions on $[a,b]$. Also, since every continuous function on a bounded closed interval has a maximum, the norm $\|\cdot\|_\infty$ on $C[a,b]$ has the form

$$\|f\|_\infty = \max\{|f(t)| : t \in [a,b]\}. \quad ☐$$

The following example is important in relation to Example 1.15.

1.16 Example. $(C[a,b], \|\cdot\|_1)$

$C[a,b]$ is a linear space with norm $\|\cdot\|_1$ defined by

$$\|f\|_1 = \int_a^b |f(t)| \, dt.$$

We call this the *integral norm* for $C[a,b]$. The only norm property which provides any difficulty to verify is that part of property (ii) which states that

$$\|f\|_1 = 0 \text{ implies that } f = 0.$$

This follows contrapositively from the following property of continuous functions:

If f is a positive continuous function on $[a,b]$ where there exists a $c \in [a,b]$ such that $f(c) > 0$ then $\int_a^b f(t) \, dt > 0$. $\square$

1.17 Example. ℓ_2-space

An important generalisation of Euclidean n-space (or Unitary n-space) which retains much of the special structure of such spaces is real (or complex) *Hilbert sequence space* denoted by ℓ_2. The set ℓ_2 whose elements are sequences of scalars $x \equiv \{\lambda_1, \lambda_2, \ldots, \lambda_n, \ldots\}$ such that $\Sigma |\lambda_n|^2$ is convergent, is a linear space under pointwise definition of the linear space operations and is a normed linear space with norm $\|\cdot\|_2$ defined by

$$\|x\|_2 = \sqrt{\left(\sum_{k=1}^{\infty} |\lambda_k|^2 \right)}.$$

We can verify simultaneously that ℓ_2 is a linear space and that $\|\cdot\|_2$ is a norm for ℓ_2. The only norm property which provides difficulty to verify is the triangle inequality.

Proof of the triangle inequality.

Given any $x \equiv \{\lambda_1, \lambda_2, \ldots, \lambda_n, \ldots\}$ and $y \equiv \{\mu_1, \mu_2, \ldots, \mu_n, \ldots\}$ in ℓ_2 we have from the triangle inequality for Euclidean n-space (Unitary n-space) that, for any $n \in \mathbb{N}$,

$$\sqrt{(\sum_{k=1}^{n} |\lambda_k + \mu_k|^2)} \leq \sqrt{(\sum_{k=1}^{n} |\lambda_k|^2)} + \sqrt{(\sum_{k=1}^{n} |\mu_k|^2)}$$

$$\leq \sqrt{(\sum_{k=1}^{\infty} |\lambda_k|^2)} + \sqrt{(\sum_{k=1}^{\infty} |\mu_k|^2)}$$

$$< \infty$$

So $\sum |\lambda_n + \mu_n|^2$ is convergent. Consequently $x + y \in \ell_2$ and also

$$\|x+y\|_2 \leq \|x\|_2 + \|y\|_2. \quad \Box$$

1.18 Example. ℓ_p^n-space, $(1 \leq p \leq \infty)$

In Examples 1.6, 1.8 and 1.9 we defined norms $\|\cdot\|_2$, $\|\cdot\|_1$ and $\|\cdot\|_\infty$ on the linear space $\mathbb{R}^n$ (or $\mathbb{C}^n$). These can be considered particular cases of a general class of norms on $\mathbb{R}^n$ (or $\mathbb{C}^n$). For any given $1 \leq p < \infty$ we define the norm $\|\cdot\|_p$ as follows:
For $x \equiv (\lambda_1, \lambda_2, \ldots, \lambda_n)$,

$$\|x\|_p = (\sum_{k=1}^{n} |\lambda_k|^p)^{\frac{1}{p}}$$

We call this a *p-norm* and refer to $\mathbb{R}^n$ (or $\mathbb{C}^n$) with this norm as real (or complex) ℓ_p^n-*space*.

The only norm property which provides any difficulty to verify is again the triangle inequality. The proof of this inequality, when $1 < p < \infty$, is derived from a generalisation of the Cauchy-Schwarz inequality.

1.19 Lemma. The Hölder inequality

In the linear space $\mathbb{R}^n$ (or $\mathbb{C}$), given $1 < p < \infty$, for any $x \equiv (\lambda_1, \lambda_2, \ldots, \lambda_n)$ *and any* $y \equiv (\mu_1, \mu_2, \ldots, \mu_n)$,

$$\sum_{k=1}^{n} |\lambda_k \mu_k| \leq (\sum_{k=1}^{n} |\lambda_k|^p)^{\frac{1}{p}} (\sum_{k=1}^{n} |\mu_k|^q)^{\frac{1}{q}}$$

where $\frac{1}{p} + \frac{1}{q} = 1$.

The proof is derived as for the Cauchy-Schwarz inequality but from the following more general elementary inequality:
For any positive real numbers a and b

$$ab \leq \frac{a^p}{p} + \frac{b^q}{q} \quad \text{for any given } 1 < p < \infty \text{ and } \frac{1}{p} + \frac{1}{q} = 1;$$

(see A.J. White *Real Analysis: an introduction*, p.40). □

Proof of the triangle inequality in 1.18.

For non-zero $x + y$,

$$\sum_{k=1}^{n} |\lambda_k + \mu_k|^p \leq \sum_{k=1}^{n} |\lambda_k + \mu_k|^{p-1} |\lambda_k| + \sum_{k=1}^{n} |\lambda_k + \mu_k|^{p-1} |\mu_k|$$

$$\leq \left(\sum_{k=1}^{n} |\lambda_k + \mu_k|^{(p-1)q} \right)^{\frac{1}{q}} \left(\sum_{k=1}^{n} |\lambda_k|^p \right)^{\frac{1}{p}}$$

$$+ \left(\sum_{k=1}^{n} |\lambda_k + \mu_k|^{(p-1)q} \right)^{\frac{1}{q}} \left(\sum_{k=1}^{n} |\mu_k|^p \right)^{\frac{1}{p}}$$

by the Hölder inequality.

But $(p-1)q = p$ and dividing by $\left(\sum_{k=1}^{n} |\lambda_k + \mu_k|^p \right)^{\frac{1}{q}}$ we have the triangle inequality. □

At this point we should show that Example 1.9 can be included in this class as the case $p = \infty$; this also provides a reason for the notation $\| \cdot \|_\infty$ for the supremum norm.

__1.20 Lemma.__ *In the linear space* $\mathbb{R}^n$ *(or* $\mathbb{C}^n$*), for any given*
$x \equiv (\lambda_1, \lambda_2, \ldots, \lambda_n)$,

$$\max \{ |\lambda_k| : k \in \{1, 2, \ldots, n\} \} = \lim_{p \to \infty} \left(\sum_{k=1}^{n} |\lambda_k|^p \right)^{\frac{1}{p}}$$

__Proof.__ For all positive real numbers a and b where $a \geq b$ and $p \geq 1$,

$$a \leq (a^p + b^p)^{\frac{1}{p}} = a \left(1 + \left(\frac{b}{a} \right)^p \right)^{\frac{1}{p}}$$

$$\leq a \cdot 2^{\frac{1}{p}} \to a \quad \text{as} \quad p \to \infty.$$

Putting a = $|\lambda_1|$ and b = $|\lambda_2|$ we have the proof for the case n = 2. It is easily seen how the proof is extended to the general case. $\square$

We sometimes refer to $\mathbb{R}^n$ (or $\mathbb{C}^n$) with norm $\|\cdot\|_\infty$ as real (or complex) ℓ_∞^n-space.

1.21 Example. ℓ_p-space, (1 $\leq$ p $\leq$ ∞)

Just as we saw that Euclidean n-space (Unitary n-space) can be generalised to Hilbert sequence space, so the same sort of generalisation can be carried out for ℓ_p^n-spaces. Given 1 $\leq$ p < ∞, the set ℓ_p whose elements are sequences of scalars, x $\equiv \{\lambda_1, \lambda_2, \ldots, \lambda_n, \ldots\}$ such that $\Sigma|\lambda_n|^p$ is convergent, is a linear space under pointwise definition of the linear space operations and is a normed linear space with norm $\|\cdot\|_p$ defined by

$$\|x\|_p = (\sum_{k=1}^{\infty} |\lambda_k|^p)^{\frac{1}{p}} .$$

Again this is called a *p-norm* and we refer to this normed linear space as real (or complex), ℓ_p-*space*.

As with Hilbert sequence space ℓ_2 we can verify simultaneously that ℓ_p is a linear space and that $\|\cdot\|_p$ is a norm for ℓ_p. Again the only norm property which provides difficulty to verify is the triangle inequality, but it follows from the corresponding triangle inequality for ℓ_p^n-space using an argument similar to that used in the proof of the triangle inequality for Hilbert sequence space ℓ_2. $\square$

We notice that the Lemma 1.20 can be generalised so that Example 1.11 can be included as the case p = ∞; this justifies the alternative notation, ℓ_∞ for m.

1.22 Lemma. *In the linear space* m, *for any* x $\equiv \{\lambda_1, \lambda_2, \ldots, \lambda_n, \ldots\}$,

$$\sup\{|\lambda_k| : k \in \mathbb{N}\} = \lim_{p \to \infty}(\sum_{k=1}^{\infty} |\lambda_k|^p)^{\frac{1}{p}}$$

Proof. From Lemma 1.20 we have for any $n \in \mathbb{N}$,

$$\lim_{p \to \infty} \left(\sum_{k=1}^{n} |\lambda_k|^p \right)^{\frac{1}{p}} = \max\{|\lambda_k| : k \in \{1,2,\ldots,n\}\} \leq \sup\{|\lambda_k| : k \in \mathbb{N}\}$$

But also for each $k \in \mathbb{N}$,

$$|\lambda_k| \leq \lim_{p \to \infty} \left(\sum_{k=1}^{\infty} |\lambda_k|^p \right)^{\frac{1}{p}}$$

so

$$\sup\{|\lambda_k| : k \in \mathbb{N}\} \leq \lim_{p \to \infty} \left(\sum_{k=1}^{\infty} |\lambda_k|^p \right)^{\frac{1}{p}} . \quad \square$$

1.23 Remark. We should notice that the examples of normed linear spaces which we have so far introduced fall into three classes depending on the type of linear space.

The *co-ordinate spaces* are those where the linear space is $\mathbb{R}^n$ (or $\mathbb{C}^n$) and this class includes the ℓ_p^n-spaces ($1 \leq p \leq \infty$).

The *sequence spaces* are those where the linear space is some linear sub-space of sequences and this class includes the ℓ_p-spaces ($1 \leq p \leq \infty$).

The *function spaces* are those where the linear space is some linear sub-space of real (or complex) functions defined on some non-empty set.

In fact all of these are function spaces because the elements of a co-ordinate space are functions on a finite set of natural numbers and the elements of a sequence space are functions on the natural numbers. $\square$

1.24 Remark. Notice that the same notation has been used to denote the norm in different example spaces. For example, $\|\cdot\|_1$ is used for the co-ordinate space ℓ_1^n, for the sequence space ℓ_1 and the function space $(C[a,b], \|\cdot\|_1)$. This has its justification in that the integral is a generalised sum and it is possible to define a meaningful elementary type of integral for functions whose domain is a subset of the natural numbers so that the norms of such functions have integral form.

We also used $\|\cdot\|_2$ to denote the norm in Euclidean (Unitary) space and Hilbert sequence space ℓ_2. The generalisation for function spaces is particularly useful. $\square$

1.25 Example. $(C[a,b], \|\cdot\|_2)$

$C[a,b]$ is a linear space with norm $\|\cdot\|_2$ defined by

$$\|f\|_2 = \left(\int_a^b |f(t)|^2 dt \right)^{\frac{1}{2}} .$$

We call this the *mean square norm* for $C[a,b]$.

The proof of norm property (ii) follows closely that for the $\|\cdot\|_1$ norm in Example 1.16. The proof of the triangle inequality is derived from the integral form of the Cauchy-Schwarz inequality.

1.26 Lemma. *For any* $f,g \in C[a,b]$,

$$\int_a^b |fg(t)|\,dt \leq \sqrt{\left(\int_a^b |f(t)|^2 dt\right)} \sqrt{\left(\int_a^b |g(t)|^2 dt\right)}.$$

Proof. For all positive real numbers a and b

$$2ab \leq a^2 + b^2.$$

Given non-zero f and g, for any $t \in [a,b]$, put

$$a \equiv \frac{|f(t)|}{\sqrt{\left(\int_a^b |f(t)|^2 dt\right)}} \quad \text{and} \quad b \equiv \frac{|g(t)|}{\sqrt{\left(\int_a^b |g(t)|^2 dt\right)}}$$

and integrating the consequent inequalities we have the result. □

1.27 Remark. We saw in Remark 1.4 that in any normed linear space the norm generates a natural metric for the space. In a linear space X which has a metric d, if d is generated by a norm $\|\cdot\|$ then it has the form

$$\|x\| = d(x,0) \quad \text{for all } x \in X.$$

However, it should be noted that not every linear space which is also a metric space has a norm which generates the metric. □

1.28 Example. For any non-trivial linear space there is no norm which generates the discrete metric. This can be shown by examining norm property (iii):
For $x \neq 0$, $d(x,0) = 1$ and so for $|\lambda| \neq 0$ or 1,

$$|\lambda| d(x,0) = |\lambda| \neq 1 = d(\lambda x,0). \quad \Box$$

We will find many linear spaces where there is a real function on the space which has all the defining norm properties except property (ii).

1.29 Definition. Given a linear space X over $\mathbb{R}$ (or $\mathbb{C}$) a *semi-norm* p for X is a real function on X which satisfies all the norm properties except (ii) and instead of (ii) satisfies:

(ii)' $p(x) = 0$ if $x = 0$.

A linear space X with a semi-norm p for X is denoted by (X,p) and is called a *semi-normed linear space*.

A semi-norm allows the possibility that for some $x \neq 0$, $p(x) = 0$ and we shall see that this is a great disadvantage indeed.

1.30 Remark. The function $e : X \times X \to \mathbb{R}$ defined by

$$e(x,y) = p(x-y)$$

has all the properties of a metric except (ii) and instead of (ii) satisfies

(ii)' $e(x,y) = 0$ if $x = y$.

Such a function is called a *semi-metric* for X and allows the possibility that for some $x \neq y$, $e(x,y) = 0$. $\square$

1.31 Example. $(R[a,b], p_1)$

The set $R[a,b]$ of Riemann integrable functions on $[a,b]$ is a linear subspace of $B[a,b]$. The real function p_1 defined on $R[a,b]$ by

$$p_1(f) = \int_a^b |f(t)| dt$$

is a semi-norm for $R[a,b]$. Now $C[a,b]$ is a linear subspace of $R[a,b]$ and although $p_1|_{C[a,b]} = \|\cdot\|_1$ as we had in Example 1.16, it is possible to have a non-continuous Riemann integrable function f on $[a,b]$ where $p_1(f) = 0$ but $f \neq 0$. $\square$

Although the lack of the full norm property (ii) upsets our intuitive notion of distance, semi-normed linear spaces such as that in Example 1.31 do have considerable interest. Fortunately they can be transformed into normed linear spaces in a natural way.

1.32 Theorem. *Given a semi-normed linear space* (X,p), *the quotient space* $X/\ker p$ *is a linear space with linear space operations defined by*

$$[x] + [y] = [x+y] \quad \textit{for all } x,y \in X$$
$$\lambda[x] = [\lambda x] \quad \textit{for all } x \in X \textit{ and scalars } \lambda,$$

and p *generates a norm* $\|\cdot\|$ *on* $X/\ker p$ *defined by*

$$\|[x]\| = p(x) \quad \textit{for any } x \in [x].$$

<u>Sketch of the proof.</u> From Definition 1.29 it is clear that $\ker p$ is a linear subspace of X. The main difficulty is in showing that the linear space operations and the norm $\|\cdot\|$ are well defined. But then $\|\cdot\|$ has all the properties of a semi-norm.
If $\|[x]\| = 0$ then $p(x) = 0$ for all $x + \ker p$, so $x \in \ker p$. That is, $\|[x]\| = 0$ implies that $[x] = [0]$ and then $\|\cdot\|$ is a norm for $X/\ker p$. $\square$

<u>1.33 Example.</u> In Example 1.31,

$$\ker p_1 = \left\{ f \in R[a,b] : \int_a^b |f(t)|\,dt = 0 \right\}$$

and the elements of the linear space $R[a,b]/\ker p_1$ are cosets of Riemann integrable functions.
Given $f \in R[a,b]$,

$$[f] = \left\{ g \in R[a,b] : \int_a^b |(f-g)(t)|\,dt = 0 \right\}. \quad \square$$

1.34 Exercises.

1.   Determine whether the following real functions on $\mathbb{R}$ are norms for $\mathbb{R}$. For the real number x,

    (i) $F_1(x) = 2|x|$,

    (ii) $F_2(x) = [|x|]$, the greatest integer less than or equal to $|x|$,

    (iii) $F_3(x) = \ln|x|$.

2.   Determine whether the following real functions on $\mathbb{R}^2$ are norms for $\mathbb{R}^2$. For $x \equiv (\lambda,\mu)$,

    (i) $F_1(x) = |\lambda|$,

    (ii) $F_2(x) = |\lambda+\mu|$,

    (iii) $F_3(x) = (\sqrt{|\lambda|} + \sqrt{|\mu|})^2$.

3.   An *Elliptical norm* $\|\cdot\|_e$ for $\mathbb{R}^2$ is defined by

$$\|x\|_e = \sqrt{\frac{\lambda^2}{9} + 4\mu^2}, \quad \text{where } x \equiv (\lambda,\mu).$$

    (i)   Prove that $\|\cdot\|_e$ is a norm for $\mathbb{R}^2$ and sketch the set $\{x \in \mathbb{R}^2 : \|x\|_e = 1\}$.

    (ii)   Consider the real functions M and m defined on $\mathbb{R}^2$ by

$$M(x) = \max\{\|x\|_e, \|x\|_1\} \quad \text{and}$$
$$m(x) = \min\{\|x\|_e, \|x\|_1\},$$

where $\|\cdot\|_1$ is the norm of Example 1.8 for $\mathbb{R}^2$.
Prove that M is a norm for $\mathbb{R}^2$ but show that m is not a norm for $\mathbb{R}^2$ and sketch the sets

$$\{x \in \mathbb{R}^2 : M(x) = 1\} \quad \text{and}$$
$$\{x \in \mathbb{R}^2 : m(x) = 1\}.$$

4.   Consider the linear space $P[0,1]$ of real polynomials on $[0,1]$.
Determine whether the following real functions on $P[0,1]$ are norms for
$P[0,1]$.
For polynomial $p(t) \equiv \alpha_0 + \alpha_1 t + \ldots + \alpha_n t^n$,

    (i) $F_1(p) = \max\{|\alpha_0|, |\alpha_1|, \ldots, |\alpha_n|\}$,

    (ii) $F_2(p) = \max\{|p(t)| : t \in [0,1]\}$,

    (iii) $F_3(p) = \sqrt{\left(\displaystyle\int_0^1 p^2(t)\,dt\right)}$

5.   The *Zero Bias metric* d_0 for $\mathbb{R}$ is defined by

$$d(x,y) = 1 + |x-y| \qquad \text{if one and only one of x and y}$$
$$\text{is strictly positive}$$
$$= |x-y| \qquad \text{otherwise}$$

    (i)   Prove that d_0 is a metric for $\mathbb{R}$.

    (ii)   Prove that there is no norm for $\mathbb{R}$ which generates d_0.

6.   The *Post Office metric* d_p for $\mathbb{R}^2$ is defined by

$$d_p(x,y) = \|x\|_2 + \|y\|_2 \qquad \text{when } x \neq y$$
$$= 0 \qquad \text{when } x = y$$

where $\|\cdot\|_2$ is the Euclidean norm for $\mathbb{R}^2$.

    (i)   Prove that d_p is a metric for $\mathbb{R}^2$.

    (ii)   Prove that there is no norm for $\mathbb{R}^2$ which generates d_p.

7.   Given a normed linear space $(X, \|\cdot\|)$, the *Radar Screen metric*
d_r for X generated by $\|\cdot\|$ is defined by

$$d_r(x,y) = \min\{1, \|x-y\|\}.$$

    (i)   Prove that d_r is a metric for X.

    (ii)   Prove that there is no norm for X which generates d_r.

8. (i)   Prove that for all $x, y \in \mathbb{R}$,

$$\big| |x| - |y| \big| \leq |x-y|.$$

(ii)   Given a normed linear space $(X, \|\cdot\|)$, prove that for all $x, y \in X$,

$$\big| \|x\| - \|y\| \big| \leq \|x-y\|.$$

(iii)   Given a metric space (X, d), prove that for all $x, y, z \in X$,

$$\big| d(x, z) - d(y, z) \big| \leq d(x, y).$$

9.   Given a non-empty set X and a function $d : X \times X \rightarrow \mathbb{R}$ with properties

(i) $d(x, y) = 0$ if and only if $x = y$,

(ii) $d(x, y) \leq d(x, z) + d(y, z)$.

Prove that d is a metric for X.
(The four defining properties of a metric can be derived from these two.)

10.   Denote by

E_0 the set of sequences of real numbers each of which has only a finite number of non-zero terms,

c_0 the set of sequences of real numbers which converge to 0,

c the set of convergent sequences of real numbers,

ℓ_1 the set of sequences of real numbers, the series of whose terms is absolutely convergent,

ℓ_2 the set of sequences of real numbers, the series of the squares of whose terms is convergent.

10. (i)   Establish an ordering of these sequence spaces as normed
           linear subspaces of $(m, \|\cdot\|_\infty)$ discussed in Example 1.11.

     (ii) Determine those linear subspaces which admit a Hilbert
           sequence space norm $\|\cdot\|_2$ as defined in Example 1.17.

11. (i)   Given metric spaces (X,d) and (Y,d'), the *product metric* d_π
           for the Cartesian product space $X \times Y$ is defined by

           $$d_\pi\big((x,y),(x',y')\big) = \max\{d(x,x'),d'(y,y')\}$$

           Prove that the product metric is indeed a metric for $X \times Y$ and
           that $d_\pi\big|_{X \times 0} = d$ and $d_\pi\big|_{0 \times Y} = d'$.

     (ii) Given normed linear spaces $(X, \|\cdot\|)$ and $(Y, \|\cdot\|')$ over the same
           scalar field, the Cartesian product $X \times Y$ is a linear space over
           the same scalar field under co-ordinatewise definition of
           linear operations;   that is,

           $$(x,y) + (x',y') = (x+x',y+y')$$
           $$\lambda(x,y) = (\lambda x, \lambda y) \qquad \text{for all } (x,y),(x',y') \in X \times Y$$
           $$\text{and scalars } \lambda.$$

           The *product norm* $\|\cdot\|_\pi$ for $X \times Y$ is defined by

           $$\|(x,y)\|_\pi = \max\{\|x\|, \|y\|'\}.$$

           Prove that the product norm is indeed a norm for $X \times Y$ and that
           $\|\cdot\|_\pi\big|_{X \times 0} = \|\cdot\|$ and $\|\cdot\|_\pi\big|_{0 \times Y} = \|\cdot\|'$ and that the product norm
           generates the product metric.

12.   Consider the linear space $R[a,b]$ of Riemann integrable functions on $[a,b]$.

(i)   For any given $1 \leqslant p < \infty$, prove that the real function

$$p_p(f) = \left(\int_a^b |f(t)|^p \right)^{\frac{1}{p}}$$

is a semi-norm for $R[a,b]$.

(ii)   Prove that $C[a,b]$ the linear space of continuous functions on $[a,b]$ is a normed linear subspace of $(R[a,b], p_p)$ for any given $1 \leqslant p < \infty$.

(iii)   Prove that $\ker p_p = \ker p_1$ for all $1 < p < \infty$ and so $R[a,b]/\ker p_p$ is the same linear space for all $1 \leqslant p < \infty$.

2. BALLS AND BOUNDEDNESS

To develop a feel for a metric or normed linear space we need to explore the geometry associated with certain fundamental sets in the space related to the metric or norm functions.

2.1 Definition. In a metric space (X,d), given $x_0 \in X$ and $r > 0$, the set

$$S(x_0;r) \equiv \{x \in X : d(x_0,x) = r\}$$

is called the *sphere centre* x_0 *and radius* r, the set

$$B[x_0;r] \equiv \{x \in X : d(x_0,x) \leq r\}$$

is called the *closed ball centre* x_0 *and radius* r, and the set

$$B(x_0;r) \equiv \{x \in X : d(x_0,x) < r\}$$

is called the *open ball centre* x_0 *and radius* r.

The Euclidean motivation for drawing attention to such sets is clear enough. However, in some metric spaces the shape of such sets hardly accords with our Euclidean intuition.

2.2 Example. In any discrete metric space (X,d), (Example 1.2), given $x_0 \in X$

$$S(x_0;1) \equiv \{x \in X : d(x_0,x) = 1\} = X \backslash \{x_0\},$$

but given $r > 0$, $r \neq 1$,

$$S(x_0;r) = \{x \in X : d(x_0,x) = r\} = \phi.$$

Also for $r > 1$,

$$B[x_0;r] = B(x_0;r) = B[x_0;1] = X$$

and for $0 < r < 1$,

$$B[x_0;r] = B(x_0;r) = B(x_0;1) = \{x_0\}. \quad \square$$

In a normed linear space the linear structure brings some orderliness.

2.3 Definition. In a normed linear space $(X, \|\cdot\|)$, the set

$$S(\underline{0};1) \equiv \{x \in X : \|x\| = 1\}$$

is called the *unit sphere*, the set

$$B[\underline{0};1] \equiv \{x \in X : \|x\| \leqslant 1\}$$

is called the *closed unit ball* and the set

$$B(\underline{0};1) \equiv \{x \in X : \|x\| < 1\}$$

is called the *open unit ball.*

2.4 Theorem. *In a normed linear space* $(X, \|\cdot\|)$, *given* $x_0 \in X$ *and* $r > 0$,

$$S(x_0;r) = x_0 + rS(0;1);$$

(that is, every sphere of the space is a translate of a strictly positive multiple of the unit sphere).

Proof.
$$S(x_0;r) \equiv \{x \in X : \|x - x_0\| = r\}.$$

Putting $y \equiv x - x_0$,

$$S(x_0;r) = \{x_0 + y \in X : \|y\| = r\}$$
$$= x_0 + S(0;r)$$

But $S(\underline{0};r) \equiv \{x \in X : \|x\| = r\}.$

Putting $y \equiv \frac{x}{r}$,

$$S(\underline{0};r) = \{ry \in X : \|y\| = 1\}$$
$$= rS(\underline{0};1) . \square$$

2.5 Remark. This is not generally true for metrics on an underlying linear space as can be seen from an examination of $\mathbb{R}^2$ with the Post Office metric d_p, (Exercise 1.34.6). Theorem 2.4 provides a geometrical hint to the solution of Exercise 1.34.6(ii). $\square$

We now consider the shape of the unit sphere in several co-ordinate space examples.

2.6 Examples.

(i)   In $(\mathbb{R}^2, \|\cdot\|_2)$, (Example 1.6),

$$S\big((0,0);1\big) = \{(\lambda,\mu) : \lambda^2 + \mu^2 = 1\}$$

(ii)   In $(\mathbb{R}^2, \|\cdot\|_1)$, (Example 1.8),

$$S\big((0,0);1\big) = \{(\lambda,\mu) : |\lambda| + |\mu| = 1\}.$$

(iii)   In $(\mathbb{R}^2, \|\cdot\|_\infty)$, (Example 1.9),

$$S\big((0,0);1\big) = \big\{(\lambda,\mu) : \max\{|\lambda|, |\mu|\} = 1\big\}$$

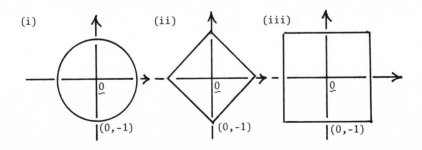

Figure 1.   The unit spheres in

(i) $(\mathbb{R}^2, \|\cdot\|_2)$ (ii) $(\mathbb{R}^2, \|\cdot\|_1)$ (iii) $(\mathbb{R}^2, \|\cdot\|_\infty)$ $\square$

With co-ordinate space examples of normed linear spaces we can assume a Euclidean background and gain some insight into the role of the unit sphere in the measurement of distance.

<u>**2.7 Remark.**</u> Consider $(\mathbb{R}^n, \|\cdot\|)$ with unit sphere $S(\underline{0};1)$. For any $x \equiv (\lambda_1, \lambda_2, \ldots, \lambda_n)$ we will determine $\|x\|$ in terms of the Euclidean norm $\|\cdot\|_2$ on $\mathbb{R}^n$:

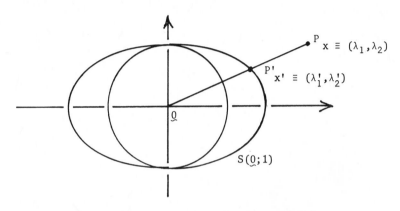

Figure 2.   The measurement of distance in $(\mathbb{R}^2, \|\cdot\|)$ described by measurement of distance in $(\mathbb{R}^2, \|\cdot\|_2)$.

Consider a ray drawn from $\underline{0}$ through point P with co-ordinates $x \equiv (\lambda_1, \lambda_2, \ldots, \lambda_n)$. Suppose this ray OP cuts $S(\underline{0};1)$ at the point P' with co-ordinates $x' \equiv (\lambda_1', \lambda_2', \ldots, \lambda_n')$. Now in the linear space $\mathbb{R}^n$ there exists an $\alpha > 0$ such that $x = \alpha x'$. We have from norm property (iii) that

$$\frac{\|x\|}{\|x'\|} = \alpha = \frac{\|x\|_2}{\|x'\|_2} .$$

But $\|x'\| = 1$ so $\|x\| = \dfrac{\|x\|_2}{\|x'\|_2}$;

that is, $\|x\| = \dfrac{|OP|}{|OP'|}$, the ratio of two Euclidean line segments. $\square$

In Euclidean space the complete symmetry of the unit sphere tells us that measurement of distance is invariant under rotation. This is not so with other normed linear spaces where the measurement of distance is dependent on the direction in which the distance is measured.

2.8 Example. In $\mathbb{R}^2$ consider the point P with co-ordinates $x \equiv (1,1)$ and rotate OP through $\frac{\pi}{4}$ to OQ where we have the point Q with co-ordinates $y \equiv (0,\sqrt{2})$. Now we can compute

$$\|x\|_2 = \|y\|_2 = \sqrt{2}$$

but $$\|x\|_\infty = 1 \quad \text{and} \quad \|y\|_\infty = \sqrt{2}$$

and $$\|x\|_1 = 2 \quad \text{and} \quad \|y\|_1 = \sqrt{2}. \;\square$$

In spaces other than co-ordinate spaces it is often difficult to visualise the spheres and balls. However, because of its many applications, we should try to gain some geometrical idea of the spheres and balls of the function space $(\mathcal{B}[a,b], \|\cdot\|_\infty)$ introduced in Example 1.12.

2.9 Example. In real $(\mathcal{B}[a,b], \|\cdot\|_\infty)$,

$$B[\underset{\sim}{0};1] \equiv \{f \in \mathcal{B}[a,b] : |f(t)| \leqslant 1 \text{ for all } t \in [a,b]\}$$

and

$$S(\underset{\sim}{0};1) \equiv \{f \in B[\underset{\sim}{0};1] : \text{there exists some } t_0 \in [a,b]$$
$$\text{such that } f(t_0) = \pm 1\}$$

So given $f_0 \in \mathcal{B}[a,b]$ we could picture

$$B[f_0;r] = f_0 + rB[0;1]$$

as those functions f which fit inside the "sleeve" around f_0,

$$f_0(t) - r \leqslant f(t) \leqslant f_0(t) + r \quad \text{for all } t \in [a,b].$$

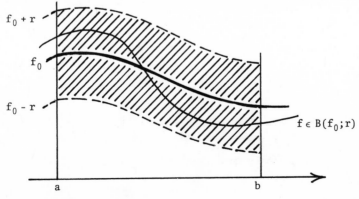

Figure 3. The function f inside the "open sleeve", $B(f_0;r)$. $\square$

It is useful to see that the norm properties give rise to corresponding geometrical properties of the unit sphere and unit balls.

2.10 Definitions. A set A in a linear space X is said to be

   (i)   *symmetric* if for any $x \in A$ we have $-x \in A$,

   (ii)   *convex* if for any $x,y \in A$ we have $\lambda x + (1-\lambda)y \in A$ for all $0 \leqslant \lambda \leqslant 1$.

2.11 Theorem. *In a normed linear space, the unit sphere and the unit balls are symmetric and the unit balls are convex.*

Proof.   (i)   Follows from norm property (iii).

          (ii)   Follows from the triangle inequality. $\square$

          Theorem 2.11 provides insight into the solution of Exercise 1.34.3(ii).

2.12 Remark. It is important to see that these geometrical properties for the balls do not hold generally for metric spaces which are also linear spaces. For example in the linear space S of all sequences of real numbers with the Fréchet metric d_f, $\underset{\sim}{B}[\underset{\sim}{0};\frac{1}{3}]$ is not convex; (see Exercise 2.33.4). $\square$

          It is instructive to examine the equivalent of the unit ball for a semi-normed linear space. Given a semi-normed linear space (X,p), it is clear as in Theorem 2.11 that $\{x \in X : p(x) \leqslant 1\}$ is symmetric and convex.

2.13 Theorem. *Given a semi-normed linear space* (X,p), p *is a norm for* X *if and only if* $\{x \in X : p(x) \leqslant 1\}$ *contains no non-trivial linear subspace of* X.

Proof. If p is a norm then for any $x \neq 0$ we have $p(x) \neq 0$ so there exists a $K > 0$ such that $p(Kx) > 1$.

          If p is not a norm then ker p is a non-trivial linear subspace of X. But

$$\{x \in X : p(x) \leqslant 1\} \supseteq \text{ker } p. \;\square$$

   The spheres and balls of a metric or normed linear space
enable us to develop our geometrical insight to run alongside the
algebraic treatment of the subject matter. As an alternative to
manipulating inequalities with the metric or norm functions we have a
geometrical interpretation involving set containment of balls. In the
following theorems we work with open balls but it is easily seen that
the results apply equally well for closed balls.

   It is important to see that in all metric spaces set containment of balls does match our Euclidean intuition.

2.14 Theorem. *In a metric space* (X,d), *given a ball* $B(x_0;r)$, *for any*
$x \in B(x_0;r)$ *we have*

$$B(x;r') \subseteq B(x_0;r) \quad \textit{for all } 0 < r' \leqslant r - d(x,x_0).$$

<u>Proof.</u> If $y \in B(x;r')$ then $d(y,x) \leqslant r'$ and

$$d(y,x_0) \leqslant d(y,x) + d(x,x_0)$$
$$< r' + d(x,x_0) \leqslant r,$$

so $y \in B(x_0;r)$.

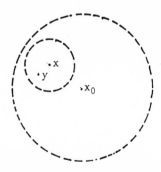

Figure 4.   The picture in $(\mathbb{R}^2, \|\cdot\|_2)$. $\square$

2.15 Remark. It is clear that Theorem 2.14 also holds for similar sets in a semi-metric space. $\square$

We often want to compare different metrics or norms on the same underlying space.

2.16 Theorem. *Given norms* $\|\cdot\|$ *and* $\|\cdot\|'$ *on a linear space* X, *for* $k > 0$, $\|x\|' \leq k\|x\|$ *for all* $x \in X$ *if and only if* $B(\underline{0};1) \subseteq B'(\underline{0};k)$.

Proof. Given that $\|x\|' \leq k\|x\|$ for all $x \in X$, if $x \in B(\underline{0};1)$, that is $\|x\| < 1$, then $\|x\|' < k$, that is $x \in B(\underline{0};k)$.

Conversely, if there exists an $x_0 \in X$ such that $\|x_0\|' > k\|x_0\|$ then

$$\left\|\frac{x_0}{\|x_0\|}\right\|' = \alpha k > k \quad \text{for some } \alpha > 1$$

so

$$\left\|\frac{x_0}{\alpha\|x_0\|}\right\|' = k.$$

Then $\dfrac{x_0}{\alpha\|x_0\|} \in B(\underline{0};1)$ but $\notin B'(\underline{0};k)$. $\square$

2.17 Example. Consider norms $\|\cdot\|_\infty$, $\|\cdot\|_2$ and $\|\cdot\|_1$ on $\mathbb{R}^n$. Now for $x \equiv (\lambda_1, \lambda_2, \ldots, \lambda_n)$,

$$\|x\|_\infty = \max\{|\lambda_k| : k \in \{1,2,\ldots,n\}\}$$

$$\leq \sqrt{\left(\sum_{k=1}^{n} |\lambda_k|^2\right)} = \|x\|_2$$

$$\leq \sum_{k=1}^{n} |\lambda_k| = \|x\|_1$$

$$\leq n \max\{|\lambda_k| : k \in \{1,2,\ldots,n\}\} = n\|x\|_\infty.$$

So $B_\infty(\underline{0};\tfrac{1}{n}) \subseteq B_1(\underline{0};1) \subseteq B_2(\underline{0};1) \subseteq B_\infty(\underline{0};1)$.

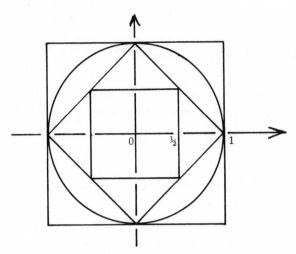

Figure 5.   In $\mathbb{R}^2$, $B_\infty[0;\tfrac{1}{2}] \subseteq B_1[0;1] \subseteq B_2[0;1] \subseteq B_\infty[0;1]$.   □

<u>2.18 Remark</u>. We should notice that homogeneity of the norm is used to
prove the converse of Theorem 2.16. This converse does not extend in
general to metric spaces. For example, in $\mathbb{R}$ with the discrete metric d
and the usual metric, there is no $k > 0$ such that $d(x,y) \leqslant k|x-y|$ for all
$x,y \in \mathbb{R}$ but $\{0\} = B_d(0;1) \subseteq B(0;k) = (-k,k)$ for all $k > 0$. □

<u>2.19 Remark</u>. Because Theorem 2.16 can be stated in terms of closed balls,
we have for a linear space X with norms $\|\cdot\|$ and $\|\cdot\|'$ that, given $x_0 \in X$
and $r > 0$,

$$B(x_0;r) \subseteq B'(x_0;r)$$

if and only if

$$B[x_0;r] \subseteq B'[x_0;r].$$

To confuse our intuition we should note that this is not true in general
for metric spaces. For example, in $\mathbb{R}$ with the discrete metric d and the
usual metric,

$$\{0\} = B_d(0;1) \subseteq B(0;1) = (-1,1),$$

but $\mathbb{R} = B_d[0;1] \nsubseteq B[0;1] = [-1,1]$.   □

Boundedness is a notion related to distance. We are familiar with the idea of a bounded set in real and complex analysis. There is an obvious generalisation to metric spaces.

<u>2.20 Definition</u>. Given a metric space (X,d), a non-empty subset A is said to be *bounded* if there exists an $x_0 \in X$ and a $k > 0$ such that

$$d(x_0,x) \leqslant k \quad \text{for all } x \in A.$$

Geometrically, $A \subseteq B[x_0;k]$; that is, A is bounded if and only if it can be contained in some ball.

<u>2.21 Remark</u>. We note that if A is bounded then for any $x' \in X$ there exists a $k' > 0$ such that

$$d(x',x) \leqslant k' \quad \text{for all } x \in A:$$

This follows from the triangle inequality

$$d(x',x) \leqslant d(x',x_0) + d(x_0,x)$$
$$\leqslant d(x',x_0) + k \qquad \text{for all } x \in A.$$

Put $k' \equiv d(x',x_0) + k.$ $\square$

Since a linear space has a preferred point $\underline{0}$ we use the following equivalent characterisation for normed linear spaces.

<u>2.22 Theorem</u>. *Given a normed linear space* $(X, \|\cdot\|)$, *a non-empty subset* A *is bounded if and only if there exists a* $k > 0$ *such that*

$$\|x\| \leqslant k \quad \text{for all } x \in A.$$

In studying boundedness in example spaces we need to be cautious of our intuition.

<u>2.23 Example</u>. In a discrete metric space (X,d), every non-empty subset A is bounded since for any $x_0 \in X$, $d(x_0,x) \leqslant 1$ for all $x \in A.$ $\square$

It is important to realise that boundedness depends on the particular metric being used.

2.24 Example. The set of natural numbers $\mathbb{N}$ is unbounded in $\mathbb{R}$ with the usual metric but is bounded in $\mathbb{R}$ with the discrete metric. $\square$

2.25 Example. Consider the linear space $C[0,1]$ and the subset of functions $A \equiv \{f_\lambda : \lambda \geq 1\}$ where for each λ,

$$f_\lambda(t) = \lambda - \lambda^2 t \qquad 0 \leq t \leq \tfrac{1}{\lambda} \left.\begin{matrix} \\ \\ \end{matrix}\right\}$$
$$= 0 \qquad\qquad \tfrac{1}{\lambda} < t \leq 1$$

Now for each $\lambda \geq 1$,

$$\|f_\lambda\|_\infty \equiv \max\{|f_\lambda(t)| : t \in [0,1]\} = \lambda,$$

so A is unbounded in $(C[0,1], \|\cdot\|_\infty)$.
But for each $\lambda \geq 1$,

$$\|f_\lambda\|_1 \equiv \int_0^1 |f_\lambda(t)| dt = \tfrac{1}{2},$$

so A is bounded in $(C[0,1], \|\cdot\|_1)$. $\square$

For bounded sets we have the associated idea of the diameter of a set.

2.26 Definition. Given a metric space (X,d) and a non-empty bounded subset A, the real number

$$\delta(A) \equiv \sup\{d(x,y) : x,y \in A\}$$

is called the *diameter of A*.

2.27 Remark. It follows from the boundedness of A that $\delta(A)$ is well defined as a real number: There exists an $x_0 \in X$ and a $k > 0$ such that

$$d(x_0,x) \leq k \qquad \text{for all } x \in A,$$

so $d(x,y) \leqslant d(x,x_0) + d(y,x_0)$

$\leqslant 2k$ for all $x,y \in A$.

Clearly $\delta(A) = 0$ if and only if A is a single point set. Also for non-empty bounded sets A and B, if $A \subseteq B$ then $\delta(A) \leqslant \delta(B)$. $\square$

We should be prepared to find situations which affront our intuitive idea of diameter.

<u>2.28 Example</u>. In any metric space (X,d), a non-empty sphere, centre x_0 and radius $r > 0$, has

$\delta\big(S(x_0;r)\big) \leqslant 2r$

but equality is not assured:
In the discrete metric space (X,d), where X has more than one element,

$S(x_0;1) = X \setminus \{x_0\}$ and

$\delta\big(S(x_0;1)\big) = 1 < 2$.

For $r > 0$, $r \neq 1$, $S(x_0;r) = \phi$ and $\delta\big(S(x_0;r)\big)$ is not defined. $\square$

In a metric space we also have the notion of distance between a point and a set and distance between two sets.

<u>2.29 Definition</u>. Given a metric space (X,d), a point $x \in X$ and a non-empty subset A of X, we define the *distance between x and A* as

$d(x,A) \equiv \inf\{d(x,y) : y \in A\}$.

Now $d(x,A)$ is well defined as a real number since, for any $y \in A$ we have $d(x,A) \leqslant d(x,y)$.

<u>2.30 Remark</u>. If $x \in A$ then $d(x,A) = 0$ but the converse is not true in general: In $\mathbb{R}$ with the usual norm, for the subset $A \equiv \{\frac{1}{n} : n \in \mathbb{N}\}$ we have $0 \notin A$ but $d(0,A) \leqslant d(0,\frac{1}{n}) = \frac{1}{n}$ for all $n \in \mathbb{N}$, so $d(0,A) = 0$. $\square$

2.31 Definition. Given a metric space (X,d) and non-empty subsets A and B of X, we define the *distance between A and B* as

$$d(A,B) \equiv \inf\{d(x,y) : x \in A, y \in B\}.$$

Again $d(A,B)$ is well defined as a real number since for all $x \in A$ and $y \in B$,

$$d(A,B) \leq d(x,y).$$

2.32 Remark. It is clear that

$$d(A,B) = \inf\{d(x,B) : x \in A\} = \inf\{d(y,A) : y \in B\}.$$

As in Remark 2.30, we see that if $A \cap B \neq \phi$ then $d(A,B) = 0$ but the converse is not true in general. $\square$

2.33 Exercises.

1. (i) Describe the sets $S(0;1)$ and $B[1;1]$ in the metric subspace $(-2,2)$ formed from $\mathbb{R}$ with
    (a) the usual norm $|\cdot|$,
    (b) the Zero Bias metric d_0, (Exercise 1.34.5),
    (c) the Radar Screen metric d_r generated by the usual norm $|\cdot|$, (Exercise 1.34.7).

   (ii) Describe the sets $S\big((1,1);2\big)$ and $B[(1,1);2]$ in the metric subspace $\{(\lambda,\mu) : \lambda,\mu \geq 0\}$ formed from $\mathbb{R}^2$ with
    (a) the Euclidean norm $\|\cdot\|_2$,
    (b) the supremum norm $\|\cdot\|_\infty$,
    (c) the Post Office metric d_p, (Exercise 1.34.6).

2. (i) Prove that if a set E is bounded in $(C[0,1],\|\cdot\|_\infty)$ then it is bounded in $(C[0,1],\|\cdot\|_1)$ but give an example to show that the converse is not true.

   (ii) Determine which of the sets

   $$E_1 \equiv \{p_n(t) = nt(1-t)^n : n \in \mathbb{N}\} \quad \text{and}$$
   $$E_2 \equiv \{p_n(t) = t + \tfrac{1}{2}t^2 + \ldots + \tfrac{1}{n}t^n : n \in \mathbb{N}\}$$

   in $P[0,1]$ the linear space of real polynomials on $[0,1]$, is bounded with respect to

(a)   the supremum norm $\|\cdot\|_\infty$ and

(b)   the integral norm $\|\cdot\|_1$.

3.   Consider a metric space (X,d). The *bounded metric* d_b for X generated by d is defined by

$$d_b(x,y) = \frac{d(x,y)}{1+d(x,y)} \ .$$

(i)   Prove that d_b is a metric for X.
        (Hint: For any positive real numbers a and b
$$\frac{a+b}{1+a+b} \leq \frac{a}{1+a} + \frac{b}{1+b} \ .)$$

(ii)   Prove that every subset A is bounded in (X,d_b) and $\delta(A) \leq 1$.

(iii)   When (X,d) is $\mathbb{R}^2$ with the Euclidean metric describe the sets $B\big((0,0);\tfrac{1}{2}\big)$ and $S\big((0,0);1\big)$.

(iv)   Prove that when X is a linear space and d is generated by a norm, then there is no norm for X which generates d_b.

4.   Consider the linear space S of all sequences of real numbers. The *Fréchet metric* d_f for S is defined as follows. For $x \equiv \{\lambda_1,\lambda_2,\ldots,\lambda_n,\ldots\}$ and $y \equiv \{\mu_1,\mu_2,\ldots,\mu_n,\ldots\}$

$$d_f(x,y) = \sum_{n=1}^\infty \frac{1}{2^n} \frac{|\lambda_n-\mu_n|}{1+|\lambda_n-\mu_n|} \ .$$

(i)   Prove that d_f is a metric for S.

(ii)   Prove that every subset A is bounded in (S,d_f) and $\delta(A) \leq 1$.

(iii)   By examining the elements

$x \equiv \{1,0,1,0,\ldots\}$ and
$y \equiv \{\tfrac{1}{2},\tfrac{1}{2},\ldots\quad\}$,   or otherwise,

show that $B[\underline{0};\tfrac{1}{3}]$ is not convex.

(iv)   Prove that there is no norm for S which generates d_f.

5.   The metric d in a metric space (X,d) is said to satisfy the
ultrametric inequality   if

$$d(x,z) \leq \max\{d(x,y),d(y,z)\}   \text{ for all } x,y,z \in X.$$

(i)   Prove that if $d(x,y) \neq d(y,z)$ then

$$d(x,z) = \max\{d(x,y),d(y,z)\}.$$

(ii)   Prove that, given $x \in X$ and $r > 0$, for any $y \in B(x;r)$ we have

$$B(y;r) = B(x;r).$$

(iii)   Prove that if two balls intersect then one is contained in the
other.

(iv)   Prove that if $B(x;r)$ and $B(y;r)$, $x \neq y$, are contained in
$B[z;r]$ then

$$d\big(B(x;r),B(y;r)\big) = r.$$

(v)   Consider the linear space S of all sequences of real numbers.
For $x \equiv \{\lambda_1,\lambda_2,\ldots,\lambda_n,\ldots\}$ and $y \equiv \{\mu_1,\mu_2,\ldots,\mu_n,\ldots\}$
denote by $k(x,y)$ the smallest integer n such that $\lambda_n \neq \mu_n$.
Define

$$d(x,y) = \frac{1}{k(x,y)}   \text{if } x \neq y$$
$$= 0   \text{if } x = y$$

Prove that d is a metric on S which has the ultrametric
inequality.

6. (i)   Given a metric space (X,d), prove that for any non-empty subset
A and $x,y \in X$,

$$d(x,A) \leq d(x,y) + d(y,A)   \text{and}$$
$$|d(x,A)-d(y,A)| \leq d(x,y).$$

(ii)   Given a metric space (X,d) and bounded subsets A and B, prove
that

$$\delta(A \cup B) \leq \delta(A) + \delta(B) + d(A,B).$$

II. LIMIT PROCESSES

The major significance in our defining such a general concept as a metric space is that the analysis of metric spaces is almost as rich in the study of limit processes as real or complex analysis. In general, metric spaces like the complex numbers, lack an order relation like the real number system, but limit processes are mainly derived from the notion of distance.

The usefulness of developing a theory of limit processes in a general metric space setting is that it provides a unifying framework for the study of limit processes in a wide variety of apparently dissimilar situations.

3. CONVERGENCE AND COMPLETENESS

We began Chapter I using convergence of sequences in real and complex analysis to motivate our definition of a metric space. Convergence of sequences is the basic idea from which all limit processes are developed. We particularly chose our metric space axioms to enable a fruitful extension of the study of convergence of sequences to metric spaces.

A sequence is the name given to any function whose domain is the set of natural numbers $\mathbb{N}$ or a proper subset of $\mathbb{N}$. Sequences with domain a finite subset of $\mathbb{N}$ are called finite sequences as distinct from infinite sequences. In analysis our interest is in infinite sequences.

3.1 Definition. A *sequence* in a metric space (X,d) is a function f from the set of natural numbers $\mathbb{N}$ into the metric space (X,d). It is common practice to use a subscript notation for a sequence and write $f(n)$ as x_n denoting the sequence by

$$\{x_n\} \equiv x_1, x_2, \ldots, x_n, \ldots$$

Any infinite subset S of natural numbers can be represented as the range
of a sequence where the terms preserve the order of the natural number
system $\mathbb{N}$;   that is, there is an order preserving function g from $\mathbb{N}$ onto S.
So $f|_S$ can be represented as $f \circ g$ from $\mathbb{N}$ into X and we call this a
subsequence of $\{x_n\}$.   We write $f \circ g(k)$ as x_{n_k} and denote the subsequence by

$$\{x_{n_k}\} \equiv x_{n_1}, x_{n_2}, \ldots, x_{n_k}, \ldots$$

3.2 Definition.   A sequence $\{x_n\}$ in a metric space (X,d) is said to
converge to $x \in X$ if, given $\varepsilon > 0$ there exists a $\nu \in \mathbb{N}$ such that

$$d(x_n, x) < \varepsilon \quad \text{for all } n > \nu;$$

geometrically, $x_n \in B(x;\varepsilon)$ for all $n > \nu$.   We say that $\{x_n\}$ is a
convergent sequence with *limit* x and we write $d(x_n, x) \to 0$ or simply
$\lim x_n = x$ or $x_n \to x$ if the metric defining the convergence is clearly
understood from the context.

3.3 Definition.   A mapping T of a set X into a metric space (Y,d) is said
to be *bounded* if the range T(X) is a bounded set in (Y,d).
So a sequence $\{x_n\}$ in a metric space (X,d) is bounded if and only if the
range of the sequence $f(\mathbb{N}) \equiv \{x_n : n \in \mathbb{N}\}$ is a bounded set in (X,d).

We would expect that convergent sequences in metric spaces
satisfy the following elementary properties.

3.4 Theorem.   *Every convergent sequence* $\{x_n\}$ *in a metric space* (X,d)

> *(i)   has a unique limit and*
> *(ii) is bounded.*

Proof.   (i)   Suppose that $\{x_n\}$ is convergent to both x and x' in X;
that is, given $\varepsilon > 0$ there exist $\nu_1, \nu_2 \in \mathbb{N}$ such that

$$d(x_n, x) < \varepsilon \quad \text{for all } n > \nu_1 \quad \text{and}$$
$$d(x_n, x') < \varepsilon \quad \text{for all } n > \nu_2.$$

But then by the triangle inequality

$$d(x,x') \leq d(x_n,x) + d(x_n,x')$$
$$< 2\varepsilon \quad \text{for all } n > \max(\nu_1,\nu_2).$$

Therefore, by metric property (ii), x = x'.

(ii) Given that $\{x_n\}$ is convergent to $x \in X$, there exists a $\nu \in \mathbb{N}$ such that

$$d(x_n,x) < 1 \quad \text{for all } n > \nu.$$

Let $r \equiv \max\{d(x_1,x),d(x_2,x),\ldots,d(x_\nu,x),1\}$. Then $d(x_n,x) \leq r$ for all $n \in \mathbb{N}$; that is, $\{x_n\}$ is bounded. $\square$

<u>3.5 Remarks</u>.(i) It is evident from the proof of Theorem 3.4(i) that semi-metric spaces do not necessarily have the desirable uniqueness of limit property.

(ii) Sometimes this innocent result, Theorem 3.4(ii) can be used contrapositively to show very simply that a sequence is not convergent. $\square$

As we consider example spaces our intuition needs constant checking against the formal definitions.

<u>3.6 Example</u>. *In a discrete metric space* (X,d), *a convergent sequence* $\{x_n\}$ *has only a finite number of points in its range.*

<u>Proof</u>. Consider a sequence $\{x_n\}$ convergent to $x \in X$. There exists a $\nu \in \mathbb{N}$ such that

$$d(x_n,x) < 1 \quad \text{for all } n > \nu.$$

But then $d(x_n,x) = 0$ for all $n > \nu$ and $x_n = x$ for all $n > \nu$. $\square$

In a normed linear space $(X, \|\cdot\|)$, we have that a sequence $\{x_n\}$ is convergent to $x \in X$ if and only if the sequence $\{x_n - x\}$ is convergent to $\underline{0}$. So any examination of convergence in general can be reduced to convergence to the zero vector $\underset{\sim}{0}$ with obvious advantages in calculation.

3.7 Example. *In Euclidean m-space* $(\mathbb{R}^m, \|\cdot\|_2)$, *a sequence* $\{x_n\}$ *is convergent to* x, *where*

$$x_n \equiv (\lambda_1^n, \lambda_2^n, \ldots, \lambda_m^n) \quad and$$
$$x \equiv (\lambda_1, \lambda_2, \ldots, \lambda_m),$$

if and only if each co-ordinate sequence $\{\lambda_k^n\}$ *is convergent to* λ_k *for* $k \in \{1,2,\ldots,m\}$; (that is, in Euclidean space, convergence is equivalent to co-ordinatewise convergence).

Proof. It is sufficient to consider the case when $x = \underset{\sim}{0}$.

Suppose that $x_n \to \underset{\sim}{0}$. Then given $\varepsilon > 0$ there exists a $\nu \in \mathbb{N}$

such that $\|x_n\|_2 < \varepsilon$ for all $n > \nu$;

that is, $\sqrt{(\sum_{k=1}^{m} |\lambda_k^n|^2)} < \varepsilon$ for all $n > \nu$

But this implies that, for each $k \in \{1,2,\ldots,m\}$,

$$|\lambda_k^n| < \varepsilon \quad \text{for all } n > \nu;$$

that is, for each $k \in \{1,2,\ldots,m\}$, we have $\{\lambda_k^n\}$ is convergent to 0.

Conversely, suppose that $\{\lambda_k^n\}$ is convergent to 0 for each $k \in \{1,2,\ldots,m\}$. Then given $\varepsilon > 0$ there exists a $\nu_k \in \mathbb{N}$ for each $k \in \{1,2,\ldots,m\}$ such that

$$|\lambda_k^n| < \varepsilon \quad \text{for all } n > \nu_k.$$

But then $\|x_n\|_2 = \sqrt{(\sum_{k=1}^{m} |\lambda_k^n|^2)}$

$< \sqrt{m}\varepsilon$ for all $n > \max\{\nu_k : k \in \{1,2,\ldots,m\}\}$;

that is, $x_n \to \underset{\sim}{0}$. $\square$

In the study of normed linear spaces those results which establish a causal link between the algebra and the analysis structures are of particular significance. The generalisation of this Euclidean space property to all finite dimensional normed linear spaces is just such a result. The particular fascination of this generalisation is that it establishes a convergence characterisation (an analysis property) simply as a consequence of the finite dimensionality of the space (an algebra property).

The proof of the Euclidean case uses inequalities derived from the norm formula. In the generalisation we overcome the lack of such a formula by using the local compactness of the real number system; that is, the property that every bounded sequence has a convergence subsequence.

3.8 Theorem. *In a finite dimensional normed linear space* $(X_m, \|\cdot\|)$ *with basis* $(e_1, e_2, \ldots, e_m)$, *a sequence* $\{x_n\}$ *is convergent to x if and only if each co-ordinate sequence* $\{\lambda_k^n\}$ *is convergent to* λ_k *for* $k \in \{1, 2, \ldots, m\}$ *where*

$$x_1 \equiv \lambda_1^1 e_1 + \lambda_2^1 e_2 + \ldots + \lambda_m^1 e_m$$
$$x_2 \equiv \lambda_1^2 e_1 + \lambda_2^2 e_2 + \ldots + \lambda_m^2 e_m$$
$$\cdot \quad \cdot \quad \cdot$$
$$x_n \equiv \lambda_1^n e_1 + \lambda_2^n e_2 + \ldots + \lambda_m^n e_m$$
$$\cdot \quad \cdot \quad \cdot$$
$$x \equiv \lambda_1 e_1 + \lambda_2 e_2 + \ldots + \lambda_m e_m$$

(that is, in any finite dimensional normed linear space convergence is equivalent to co-ordinatewise convergence).

Proof. It is sufficient to consider the case where $x = \underline{0}$.

Suppose that each co-ordinate sequence $\{\lambda_k^n\}$ is convergent to 0. Then given $\epsilon > 0$, for each $k \in \{1, 2, \ldots, m\}$ there exists a $\nu_k \in \mathbb{N}$ such that $|\lambda_k^n| < \epsilon$ for all $n > \nu_k$. Therefore,

$$|\lambda_k^n| < \epsilon \quad \text{for all } n > \nu \equiv \max\{\nu_1, \nu_2, \ldots, \nu_m\},$$
$$\text{and all } k \in \{1, 2, \ldots, m\}.$$

Now $\quad \|x_n\| \le |\lambda_1^n| \|e_1\| + |\lambda_2^n| \|e_2\| + \ldots + |\lambda_m^n| \|e_m\| \le \left(\sum_{k=1}^{m} \|e_k\| \right) \epsilon$ for all $n > \nu$;

that is, $\{x_n\}$ is convergent to $\underline{0}$.

Conversely, suppose that $\{x_n\}$ is convergent to $\underline{0}$ but for some $k_0 \in \{1, 2, \ldots, m\}$, the co-ordinate sequence $\{\lambda_{k_0}^n\}$ is not convergent to 0.

Then there exists a subsequence of $\{x_n\}$, (for convenience of notation still referred to as $\{x_n\}$), and an $r > 0$ such that $\left|\lambda^n_{k_0}\right| > r$ for all $n \in \mathbb{N}$. For each $n \in \mathbb{N}$, write $M_n \equiv \max\{\left|\lambda^n_k\right| : k \in \{1,2,\ldots,m\}\}$ and consider $y_n \equiv \dfrac{x_n}{M_n}$. Now $M_n > r$ for all $n \in \mathbb{N}$ and

$$\|y_n\| = \frac{\|x_n\|}{M_n} < \frac{1}{r}\|x_n\|,$$

so $\{y_n\}$ is also convergent to $\underset{\sim}{0}$. From $\{y_n\}$ we choose a subsequence which converges co-ordinatewise but to a non-zero element:

Notice that for each $n \in \mathbb{N}$ the co-ordinates of y_n lie between -1 and $+1$ and there exists at least one equal to -1 or $+1$. So then each co-ordinate sequence of $\{y_n\}$ is bounded and at least one of these has a convergent subsequence which consists of terms -1 or $+1$. Consider the corresponding subsequence of $\{y_n\}$ and its first co-ordinate sequence. By the local compactness of $\mathbb{R}$, this has a convergent subsequence. Consider now the corresponding subsequence of the second co-ordinate sequence. By the local compactness of $\mathbb{R}$, this has a convergent subsequence. Continuing this process to the mth co-ordinate sequence we obtain a subsequence $\{y_{n_\ell}\}$ which has each co-ordinate sequence convergent but one of these co-ordinate sequences is not convergent to 0. So the co-ordinate limit is not $\underset{\sim}{0}$. Using the first part of the theorem we see that this contradicts the fact that $\{y_n\}$ is convergent to $\underset{\sim}{0}$. $\square$

3.9 Remark.

Such a result holds only partially in sequence and function spaces of the type $(\mathcal{B}(X), \|\cdot\|_\infty)$.

For any sequence $\{f_n\}$, we have for any $x \in X$ that

$$|f_n(x)| \le \sup\{|f_n(x) : x \in X\}$$
$$= \|f_n\|_\infty .$$

So if $\{f_n\}$ is convergent to $\underset{\sim}{0}$ then $\{f_n(x)\}$ is convergent to 0 for each $x \in X$.

However, in $(m, \|\cdot\|_\infty)$ consider the sequence $\{x_n\}$ where

$$x_1 \equiv \{1,0,\ldots \qquad \}$$
$$x_2 \equiv \{0,1,0,\ldots \qquad \}$$
$$\cdot \ \cdot \ \cdot$$
$$x_n \equiv \{0,0,\ldots,0,1,0,\ldots\}$$
$$\text{nth place}$$
$$\cdot \ \cdot \ \cdot$$

Now for each $k \in \mathbb{N}$, $\{\lambda_k^n\}$ is convergent to 0 but $\|x_n\|_\infty = 1$ for all $n \in \mathbb{N}$ and so $\{x_n\}$ is not convergent to $\underset{\sim}{0}$.

It is not difficult to construct a similar example sequence in $(\mathcal{B}[a,b], \|\cdot\|_\infty)$; (see Exercise 3.53.3). $\square$

Nevertheless, this observation does guide us in the following way. If, in a space of this type, we have a sequence we are testing for convergence, then the logical "guess limit" to test is the co-ordinatewise or pointwise limit.

3.10 Example. In $(c_0, \|\cdot\|_\infty)$, consider the sequence

$$x_1 \equiv \{1,0,\ldots \qquad \}$$
$$x_2 \equiv \{1,\tfrac{1}{2},0,\ldots \qquad \}$$
$$\cdot \ \cdot \ \cdot$$
$$x_n \equiv \{1,\tfrac{1}{2},\ldots,\tfrac{1}{n},0,\ldots\}.$$
$$\cdot \ \cdot \ \cdot$$

Now the co-ordinatewise limit is

$$x \equiv \{1,\tfrac{1}{2},\ldots,\tfrac{1}{n},\ldots\}$$

and from Remark 3.9 this is the only possible limit. Testing this as a "guess limit" we have

$$\|x_n - x\|_\infty = \|\{0,\ldots,0,\tfrac{1}{n+1}, \tfrac{1}{n+2},\ldots\}\|_\infty$$
$$= \tfrac{1}{n+1} \to 0;$$

that is, $x_n \to x$. $\square$

3.11 Example. In $(C[0,1], \|\cdot\|_\infty)$, consider the sequence

$$f_n(t) = \frac{nt}{n+t} \ .$$

Now the pointwise limit is

$$f(t) = t;$$

(fixing t and letting n vary). From Remark 3.9 this is the only possible limit. Testing this as "guess limit" we have

$$(f_n - f)(t) \ = \ \frac{nt}{n+t} - t \ = \ \frac{-t^2}{n+t} \ .$$

Now $\qquad |(f_n - f)(t)| \ = \ \frac{t^2}{n+t} \leq \frac{1}{n+1} \qquad$ for all $t \in [0,1]$,

so $\qquad \|f_n - f\|_\infty \ \leq \ \frac{1}{n+1} \to 0 \quad$ as $\quad n \to \infty;$

that is, $\{f_n\}$ converges to f.

Such an inequality may not be obvious but we may in some cases use elementary calculus to determine, for fixed n, the maximum value of $|(f_n - f)(t)|$:

Now $\qquad (f_n' - f')(t) \ = \ \frac{-t(1+2n)}{(n+t)^2}$

$$< \ 0 \quad \text{for } t \in (0,1].$$

So $|(f_n - f)(t)|$ has a maximum value of $\frac{1}{n+1}$ at $t = 1$.

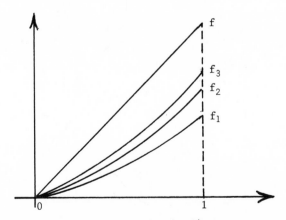

Figure 6. The functions $f_n(t) = \frac{nt}{n+t}$ converging uniformly to the function $f(t) = t$ on $[0,1]$. $\qquad\qquad$ □

Convergence in the function space $\left(B(X), \|\cdot\|_\infty\right)$ is known as *uniform convergence*. In real and complex analysis, uniform convergence has many significant applications.

It is convenient to extend the notation of uniform convergence to unbounded functions. But to do so we encounter difficulty in fitting such functions into our normed linear space structure because the supremum norm can only be defined for bounded functions. We overcome this obstacle by the following device.

3.12 Definition. A sequence of scalar functions $\{f_n\}$ on a set X is said to be *uniformly convergent* to f on X if $f_n - f \in B(X)$ for all $n \in \mathbb{N}$ and the sequence $\{f_n - f\}$ is convergent to $\underset{\sim}{0}$ in $\left(B(X), \|\cdot\|_\infty\right)$.

The significance of uniform convergence lies in the following situation:
A sequence of continuous functions may be pointwise convergent to a discontinuous function!

3.13 Example. Consider the sequence $\{f_n\}$ in the linear space $B[0,1]$

where $f_n(t) = t^n$.

The pointwise limit is

$$f(t) = 0 \quad t \in [0,1) \quad \Bigg\}$$
$$ = 1 \quad t = 1$$

However, $\|f_n - f\|_\infty = 1$ for all $n \in \mathbb{N}$, so convergence is not uniform.

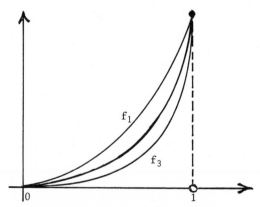

Figure 7.   The functions $f_n(t) = t^n$ converging pointwise
to a discontinuous limit function.   □

However, when convergence is uniform the situation is more as we would expect.

<u>3.14 Theorem</u>. *If a sequence of scalar functions* $\{f_n\}$ *is uniformly convergent to a function f on the interval J and for every* $n \in \mathbb{N}$, f_n *is continuous at* $t_0 \in J$ *then f is continuous at* t_0; (which implies that the limit of a uniformly convergent sequence of continuous functions is continuous).

<u>Proof</u>. The sequence $\{f_n - f\}$ is convergent to $\underline{0}$ in the normed linear space $\left(B(J), \|\cdot\|_\infty\right)$; that is, given $\varepsilon > 0$ there exists a $\nu \in \mathbb{N}$ such that

$$\|f_n - f\|_\infty < \varepsilon \quad \text{for all } n > \nu.$$

Since $f_{\nu+1}$ is continuous at t_0, there exists a $\delta > 0$ such that

$$|f_{\nu+1}(t) - f_{\nu+1}(t_0)| < \varepsilon \quad \text{when } |t - t_0| < \delta.$$

Therefore,

$$|f(t) - f(t_0)| \leq |f(t) - f_{\nu+1}(t)| + |f_{\nu+1}(t) - f_{\nu+1}(t_0)| + |f_{\nu+1}(t_0) - f(t_0)|$$

$$\leq 2\|f_{\nu+1} - f\|_\infty + |f_{\nu+1}(t) - f_{\nu+1}(t_0)|$$

$$< 3\varepsilon \quad \text{when } |t - t_0| < \delta;$$

that is, f is continuous at t_0. $\square$

<u>3.15 Remark</u>. This theorem can be used contrapositively when testing a sequence of continuous functions for uniform convergence. In Example 3.13, we can deduce from the discontinuity of the pointwise limit that the convergence is not uniform. $\square$

It should be obvious that the convergence or otherwise of a sequence depends on the particular metric being used.

<u>3.16 Example</u>. In $\mathbb{R}$, the sequence $\{\frac{1}{n}\}$ converges to 0 with respect to the usual norm, but does not converge at all with respect to the discrete metric, (see Example 3.6). □

<u>3.17 Example</u>. In the linear space $C[0,1]$, consider the sequence $\{f_n\}$ where for each $n \in \mathbb{N}$,

$$f_n(t) = t^n .$$

Now $\|f_n\|_1 = \int_0^1 |f_n(t)| dt = \dfrac{1}{n+1} \to 0$ as $n \to \infty,$

so $\{f_n\}$ converges to $\underset{\sim}{0}$ with respect to the integral norm, $\|\cdot\|_1$.
But we see from Example 3.13 that $\{f_n\}$ does not converge with respect to the supremum norm, $\|\cdot\|_\infty$. □

It is extremely useful to know when convergence is maintained with two different metrics.

<u>3.18 Definition</u>. Given a non-empty set X, metrics d and d' on X are said to be *equivalent* if for any $x \in X$ and sequence $\{x_n\}$ in X, $\{x_n\}$ is convergent to x in (X,d) if and only if $\{x_n\}$ is convergent to x in (X,d'). Given a linear space X, norms $\|\cdot\|$ and $\|\cdot\|'$ on X are said to be *equivalent* if the metrics generated by $\|\cdot\|$ and $\|\cdot\|'$ are equivalent.

<u>3.19 Remark</u>. It is clear from Example 3.16 that, on $\mathbb{R}$ the discrete metric and the usual metric are not equivalent, and from Example 3.17 that, on $C[0,1]$, the integral norm $\|\cdot\|_1$ and the supremum norm $\|\cdot\|_\infty$ are not equivalent. □

<u>3.20 Corollary to Theorem 3.8</u>. *On any given finite dimensional linear space all norms are equivalent.*

<u>3.21 Remark</u>. This fact is frequently applied in numerical analysis where we exploit our freedom to choose any norm for $\mathbb{R}^n$ which simplifies calculations. □

<u>3.22 Remark</u>. We see from Example 3.16 that Corollary 3.20 does not extend to metrics in general. □

It is particularly useful to have the following characterisation of equivalent norms.

3.23 Theorem. *Given a linear space X, norms $\|\cdot\|$ and $\|\cdot\|'$ are equivalent if and only if there exist m,M > 0 such that*

$$m\|x\| \leq \|x\|' \leq M\|x\| \quad \text{for all } x \in X.$$

Proof. Given that there exist m,M > 0 such that

$$m\|x\| \leq \|x\|' \leq M\|x\| \quad \text{for all } x \in X$$

it is evident that for any sequence $\{x_n\}$, $\|x_n\| \to 0$ if and only if $\|x_n\|' \to 0$, which implies that $\|\cdot\|$ and $\|\cdot\|'$ are equivalent.

Conversely, suppose that there does not exist an M > 0 such that

$$\|x\|' \leq M\|x\| \quad \text{for all } x \in X;$$

that is, for every $n \in \mathbb{N}$ there exists an $x_n \in X$ such that

$$\|x_n\|' > n\|x_n\|.$$

Then $\left\|\frac{x_n}{\|x_n\|'}\right\| \leq \frac{1}{n}$ but $\left\|\frac{x_n}{\|x_n\|'}\right\|' = 1$ for all $n \in \mathbb{N}$

So the sequence $\left\{\frac{x_n}{\|x_n\|'}\right\}$ is convergent to $\underset{\sim}{0}$ with respect to norm $\|\cdot\|$ but is not convergent to $\underset{\sim}{0}$ with respect to norm $\|\cdot\|'$. Symmetry of the roles of $\|\cdot\|$ and $\|\cdot\|'$ completes the proof. $\square$

3.24 Remark. Such a result holds only partially for metric spaces. It is clear that given any non-empty set X, metrics d and d' are equivalent if there exist m,M > 0 such that

$$md(x,y) \leq d'(x,y) \leq Md(x,y) \quad \text{for all } x,y \in X.$$

However, that the converse does not hold is seen from an examination of Exercise 2.33.3. $\square$

It is immediate from Theorem 3.23 that boundedness is an invariant for a linear space under equivalent norms.

3.25 Corollary. *Consider a linear space* X *with equivalent norms* $\|\cdot\|$ *and* $\|\cdot\|'$. *A subset* A *of* X *is bounded in* $(X, \|\cdot\|)$ *if and only if it is bounded in* $(X, \|\cdot\|')$.

3.26 Remark. This result does not extend to a non-empty set with equivalent metrics. See Exercise 3.53.9. □

In a normed linear space we have the following algebra of limits theorem which is analogous in part to the algebra of limits theorem of real and complex analysis.

3.27 Theorem. *In a normed linear space* $(X, \|\cdot\|)$, *given sequences* $\{x_n\}$ *and* $\{y_n\}$ *where* $x_n \to x$ *and* $y_n \to y$, *and a sequence of scalars* $\{\lambda_n\}$ *where* $\lambda_n \to \lambda$, *then*

  (i) $x_n + y_n \to x + y$ *and*
  (ii) $\lambda_n x_n \to \lambda x$.

Proof. These results follow as for real sequences using the following norm inequalities,

  (i) $\|(x_n + y_n) - (x+y)\| \le \|x_n - x\| + \|y_n - y\|$ and
  (ii) $\|\lambda_n x_n - \lambda x\| \le |\lambda_n| \|x_n - x\| + \|x\| |\lambda_n - \lambda|$. □

3.28 Remark. It follows from this theorem that the set of convergent sequences on $(X, \|\cdot\|)$ is itself a linear space. □

As in real and complex analysis we can define Cauchy sequences, but the relation between Cauchy and convergent sequences is more significant in the more general setting of metric spaces.

3.29 Definition. A sequence $\{x_n\}$ in a metric space (X, d) is said to be a *Cauchy sequence* if given $\varepsilon > 0$ there exists a $\nu \in \mathbb{N}$ such that

$$d(x_n, x_m) < \varepsilon \quad \text{for all } m, n > \nu.$$

<u>3.30 Theorem</u>. *In a metric space* (X,d),

   (i) every convergent sequence is Cauchy, and

   (ii) every Cauchy sequence is bounded.

<u>Proof</u>. (i) Consider a sequence $\{x_n\}$ convergent to $x \in X$. Then given $\varepsilon > 0$ there exists a $\nu \in \mathbb{N}$ such that

$$d(x_n,x) < \varepsilon \qquad \text{for all } n > \nu.$$

Therefore, $d(x_m,x_n) \leq d(x_n,x) + d(x_m,x)$

$$< 2\varepsilon \qquad \text{for all } m,n > \nu;$$

that is, $\{x_n\}$ is a Cauchy sequence.

  (ii) Given a Cauchy sequence $\{x_n\}$, there exists a $\nu \in \mathbb{N}$ such that

$$d(x_n,x_{\nu+1}) < 1 \qquad \text{for all } n > \nu.$$

The proof then follows that of Theorem 3.4(ii). $\square$

<u>3.31 Remark</u>. There exist metric spaces where not every Cauchy sequence is convergent. The simplest examples of such spaces are those which can be identified readily as subspaces. For example, in $\mathbb{R}$ with the usual metric, the sequence $\{\frac{1}{n}\}$ is convergent to 0, but in $(0,1)$ with the usual metric it is clear that $\{\frac{1}{n}\}$ is Cauchy but not convergent as $0 \notin (0,1)$.
In Example 3.10 we saw that the sequence

$$x_1 \equiv \{1,0,\dots \qquad\qquad \}$$
$$x_2 \equiv \{1,\tfrac{1}{2},0,\dots \qquad\quad \}$$
$$\cdot\ \cdot\ \cdot$$
$$x_n \equiv \{1,\tfrac{1}{2},\dots,\tfrac{1}{n},0,\dots\}$$
$$\cdot\ \cdot\ \cdot$$

is convergent to

$$x \equiv \{1,\tfrac{1}{2},\dots,\tfrac{1}{n},\dots\}$$

in $(c_0, \|\cdot\|_\infty)$. But in the normed linear subspace $(E_0, \|\cdot\|_\infty)$ where E_0 consists of sequences which have only a finite number of non-zero terms, $\{x_n\}$ is Cauchy but not convergent as $x \notin E_0$. $\square$

3.32 Definition. A metric space where every Cauchy sequence is convergent is called a *complete* metric space. A normed linear space which is complete as a metric space is called a *Banach* space.

3.33 Remark. It is basic to the development of real analysis that $\mathbb{R}$ with the usual norm is complete. This completeness property is deduced from the Supremum Axiom for the real number system. Once established for $\mathbb{R}$ with the usual norm it can be shown readily that $\mathbb{C}$ with the usual norm is complete. $\square$

3.34 Example. *Any discrete metric space* (X,d) *is complete.*

Proof. Given any Cauchy sequence $\{x_n\}$ in (X,d) there exists a $\nu \in \mathbb{N}$ such that

$$d(x_n, x_m) < 1 \quad \text{for all } m,n > \nu.$$

But then $d(x_n, x_{\nu+1}) < 1$ for all $n > \nu$,

and so $x_n = x_{\nu+1}$ for all $n \geqslant \nu+1$;
that is, $\{x_n\}$ converges to $x_{\nu+1}$. $\square$

3.35 Example. *Euclidean space* $(\mathbb{R}^m, \|\cdot\|_2)$ *is complete.*

Proof. Consider a Cauchy sequence $\{x_n\}$ where

$$x_n \equiv \left(\lambda_1^n, \lambda_2^n, \ldots, \lambda_m^n\right);$$

then given $\varepsilon > 0$ there exists a $\nu \in \mathbb{N}$ such that

$$\|x_k - x_\ell\|_2 = \sqrt{\left(\sum_{j=1}^m |\lambda_j^k - \lambda_j^\ell|^2\right)}$$
$$< \varepsilon \quad \text{for all } k,\ell > \nu.$$

But this implies that for each $j \in \{1,2,\ldots,m\}$,

$$|\lambda_j^k - \lambda_j^\ell| < \varepsilon \quad \text{for all } k,\ell > \nu;$$

that is, for each $j \in \{1,2,\ldots,m\}$ the sequence $\{\lambda_j^n\}$ is a Cauchy sequence in $\mathbb{R}$ with the usual norm. Since $\mathbb{R}$ with the usual norm is complete, for each $j \in \{1,2,\ldots,m\}$ there exists a λ_j such that $\{\lambda_j^n\}$ converges to λ_j. It follows from Example 3.7 that $\{x_n\}$ converges to $x \equiv (\lambda_1,\lambda_2,\ldots,\lambda_m)$. $\square$

It is significant to note that completeness is a property inherent in all finite dimensional normed linear spaces.

<u>3.36 Theorem</u>. *Every finite dimensional normed linear space* $(X_m, \|\cdot\|)$ *is complete.*

To establish the proof of this theorem we need the following lemma which is a consequence of Theorem 3.8.

<u>3.37 Lemma</u>. *In a finite dimensional normed linear space* $(X_m, \|\cdot\|)$ *with basis* $(e_1,e_2,\ldots,e_m)$, *if a sequence* $\{x_n\}$ *is Cauchy then every co-ordinate sequence* $\{\lambda_j^n\}$ *is Cauchy for* $j \in \{1,2,\ldots,m\}$ *where*

$$x_1 \equiv \lambda_1^1 e_1 + \lambda_2^1 e_2 + \ldots + \lambda_m^1 e_m$$
$$x_2 \equiv \lambda_1^2 e_1 + \lambda_2^2 e_2 + \ldots + \lambda_m^2 e_m$$
$$\cdot \cdot \cdot$$
$$x_n \equiv \lambda_1^n e_1 + \lambda_2^n e_2 + \ldots + \lambda_m^n e_m$$
$$\cdot \cdot \cdot$$

<u>Proof</u>. Suppose that there exists a co-ordinate sequence $\{\lambda_{j_0}^n\}$ which is not Cauchy; that is, there exists an $r > 0$ such that for each $k \in \mathbb{N}$ there exist $m_k, n_k > k$ where

$$\left| \lambda_{j_0}^{m_k} - \lambda_{j_0}^{n_k} \right| > r.$$

So the sequence $\{\lambda_{j_0}^{m_k} - \lambda_{j_0}^{n_k}\}$ does not converge to 0. But then from Theorem 3.8, the sequence $\{x_{m_k} - x_{n_k}\}$ does not converge to $\underset{\sim}{0}$; but this contradicts the fact that $\{x_n\}$ is Cauchy. $\square$

<u>Proof of Theorem 3.36</u>. Consider a Cauchy sequence $\{x_n\}$ in $(X_m, \|\cdot\|)$ where

$$x_1 \equiv \lambda_1^1 e_1 + \lambda_2^1 e_2 + \ldots + \lambda_m^1 e_m$$
$$x_2 \equiv \lambda_1^2 e_1 + \lambda_2^2 e_2 + \ldots + \lambda_m^2 e_m$$
$$\cdot \quad \cdot \quad \cdot$$
$$x_n \equiv \lambda_1^n e_1 + \lambda_2^n e_2 + \ldots + \lambda_m^n e_m$$
$$\cdot \quad \cdot \quad \cdot$$

From Lemma 3.37, each co-ordinate sequence $\{\lambda_j^n\}$ is Cauchy for $j \in \{1,2,\ldots,m\}$. But $\mathbb{R}$ with the usual norm is complete so for each $j \in \{1,2,\ldots,m\}$ there exists a λ_j such that $\{\lambda_j^n\}$ converges to λ_j. It follows from Theorem 3.8 that $\{x_n\}$ converges to $x \equiv \lambda_1 e_1 + \lambda_2 e_2 + \ldots + \lambda_m e_m$. $\square$

The following example contains many particular cases. We also use this example to point out the general method followed to prove completeness in other examples.

3.38 Example. *For any non-empty set* X, $(B(X), \|\cdot\|_\infty)$ *is complete.*

Proof.   (i) We begin by considering a general Cauchy sequence, we "unravel" the norm and use the completeness of the scalar field to find a candidate for the limit of the sequence:
Consider a Cauchy sequence $\{f_n\}$; then given $\varepsilon > 0$ there exists a $\nu \in \mathbb{N}$ such that

$$\|f_n - f_m\|_\infty < \varepsilon \quad \text{for all } m,n > \nu;$$

that is,

$$\sup\{\,|(f_n - f_m)(x) \; : x \in X\} < \varepsilon \quad \text{for all } m,n > \nu.$$

But then for each $x \in X$,

$$|f_n(x) - f_m(x)| < \varepsilon \quad \text{for all } m,n > \nu;$$

that is, for each $x \in X$, $\{f_n(x)\}$ is a Cauchy sequence in the scalar field. Since the scalar field is complete, for each $x \in X$ we can define a function f on X by

$$f(x) \equiv \lim f_n(x).$$

(ii) Next we show that the candidate is actually an element
of the space:

Now from Theorem 3.30(ii) we have that the sequence $\{f_n\}$ is bounded in
$(\mathcal{B}(X), \|\cdot\|_\infty)$.   So there exists a $K > 0$ such that

$$\|f_n\|_\infty < K \quad \text{for all } n \in \mathbb{N},$$

which implies that

$$|f_n(x)| < K \quad \text{for all } x \in X \text{ and } n \in \mathbb{N}.$$

Therefore, $|f(x)| \leq K$ for all $x \in X$, and so $f \in \mathcal{B}(X)$.

(iii) We finally show that the sequence converges to the
candidate limit:

We had for each $x \in X$,

$$|f_n(x) - f_m(x)| < \varepsilon \quad \text{for all } m, n > \nu.$$

Imagine n fixed and $m \to \infty$; then we have for each $x \in X$,

$$|f_n(x) - f(x)| \leq \varepsilon \quad \text{for all } n > \nu.$$

So $\|f_n - f\|_\infty = \sup\{|f_n(x) - f(x)| : x \in X\}$

$$\leq \varepsilon \quad \text{for all } n > \nu;$$

that is, $\{f_n\}$ converges to f. $\square$

3.39 Remark. Example 3.38 establishes that $(m, \|\cdot\|_\infty)$, and $(\mathcal{B}(J), \|\cdot\|_\infty)$ for
any interval J, are Banach spaces. $\square$

It is instructive to see that completeness is an invariant for
a linear space under equivalent norms.

3.40 Theorem. *Consider a linear space X with equivalent norms $\|\cdot\|$ and*
$\|\cdot\|'$. $(X, \|\cdot\|)$ is complete if and only if $(X, \|\cdot\|')$ is complete.

Proof. From Theorem 3.23 there exist k,K > 0 such that $k\|x\| \le \|x\|' \le K\|x\|$ for all $x \in X$. Suppose that $(X, \|\cdot\|)$ is complete. Consider a Cauchy sequence $\{x_n\}$ in $(X, \|\cdot\|')$. Then since

$$k\|x_n - x_m\| \le \|x_n - x_m\|' \quad \text{for all } m,n \in \mathbb{N},$$

we have that $\{x_n\}$ is a Cauchy sequence in $(X, \|\cdot\|)$. But since $(X, \|\cdot\|)$ is complete, there exists an $x \in X$ such that $\{x_n\}$ is convergent to x in $(X, \|\cdot\|)$. However,

$$\|x_n - x\|' \le K\|x_n - x\| \quad \text{for all } n \in \mathbb{N},$$

so $\{x_n\}$ is convergent to x in $(X, \|\cdot\|')$ and we conclude that $(X, \|\cdot\|')$ is complete.

The converse follows by similar argument. □

3.41 Remark. This theorem does not extend to a non-empty set with equivalent metrics. See Exercise 3.53.9. □

So far in our discussion of convergence we have confined ourselves to convergence of sequences. However, the linear structure of normed linear spaces enables us to study the convergence of series in such spaces.

3.42 Definition. Given a sequence $\{x_n\}$ in a normed linear space $(X, \|\cdot\|)$, for each $n \in \mathbb{N}$ we form the partial sum

$$s_n \equiv \sum_{k=1}^{n} x_k.$$

The *series* defined by $\{x_n\}$ in X is the sequence $\{s_n\}$ and we denote it by Σx_n.

3.43 Definition. Given a series Σx_n in a normed linear space $(X, \|\cdot\|)$, if the sequence of partial sums $\{s_n\}$ converges to s we say that the series Σx_n *converges* to s and we write

$$s \equiv \lim s_n = \lim \sum_{k=1}^{n} x_k \equiv \sum_{k=1}^{\infty} x_k.$$

Of course, not all series converge so it is important to recognise that if Σx_n does not converge then Σx_n does not represent any element in X.

It is important not to neglect the role played by completeness in the study of convergence of series. In real and complex analysis it is tacitly assumed but in this more general situation we often have to check for completeness.

3.44 Definition. A series Σx_n in a normed linear space $(X, \|\cdot\|)$ is said to be a *Cauchy series* if, given $\varepsilon > 0$ there exists a $\nu \in \mathbb{N}$ such that

$$\left\| \sum_{k=n}^{m} x_k \right\| < \varepsilon \quad \text{for all } m > n > \nu.$$

It is clear that a series Σx_n being Cauchy is equivalent to the sequence of partial sums $\{s_n\}$ being a Cauchy sequence. So the following result is immediate.

3.45 Theorem. *In a normed linear space every convergent series is Cauchy. If the normed linear space is complete then every Cauchy series is convergent.*

As in real and complex analysis we can deduce the following result.

3.46 Corollary. *In a normed linear space* $(X, \|\cdot\|)$, *if the series* Σx_n *is convergent then* $\lim \|x_n\| = 0$.

We can introduce a useful generalisation of the idea of absolute convergence.

3.47 Definition. In a normed linear space $(X, \|\cdot\|)$, a series Σx_n is said to be *absolutely convergent* if the associated real series $\Sigma \|x_n\|$ is convergent in $\mathbb{R}$ with the usual norm.

Notice that the following theorem relating absolute convergence to convergence requires completeness of the normed linear space.

3.48 Theorem. *In a Banach space* $(X, \|\cdot\|)$, *if a series* Σx_n *is absolutely convergent then it is convergent and*

$$\| \sum_{k=1}^{\infty} x_k \| \leq \sum_{k=1}^{\infty} \|x_k\|.$$

Proof. By Theorem 3.45, the series $\Sigma\|x_n\|$ is Cauchy; that is, given $\varepsilon > 0$ there exists a $\nu \in \mathbf{N}$ such that

$$\sum_{k=n}^{m} \|x_k\| < \varepsilon \quad \text{for all } m > n > \nu.$$

But $\qquad \| \sum_{k=n}^{m} x_k \| \leq \sum_{k=n}^{m} \|x_k\| \quad \text{for any } m > n \in \mathbf{N}.$

Therefore $\| \sum_{k=n}^{m} x_k \| < \varepsilon$ for all $m > n > \nu$; that is, Σx_n is Cauchy. So by Theorem 3.45, Σx_n is convergent. Now for $n = 1$ and any $m > 1$

$$\| \sum_{k=1}^{m} x_k \| \leq \sum_{k=1}^{m} \|x_k\|.$$

Therefore,

$$\| \sum_{k=1}^{m} x_k \| \leq \sum_{k=1}^{\infty} \|x_k\| \text{ for all } m > 1.$$

Now Σx_n being convergent to s means that the sequence of partial sums $\{s_n\}$ is convergent to s. Since

$$\left| \|s_n\| - \|s\| \right| \leq \|s_n - s\|$$

we have that the sequence of reals $\{\|s_n\|\}$ is convergent to $\|s\|$. Therefore,

$$\| \sum_{k=1}^{\infty} x_k \| \equiv \|s\| = \lim_{n\to\infty} \| \sum_{k=1}^{n} x_k \|$$
$$\leq \sum_{k=1}^{\infty} \|x_k\|. \quad \square$$

It can be quite a difficult matter to determine whether a series is convergent. Theorem 3.48 enables us to use techniques developed in real analysis for the more general normed linear space setting. A simple illustration is given in the following result which applies in the Banach space $(B(X), \|\cdot\|_\infty)$.

3.49 Weierstrass' M-test. *For any series of bounded functions Σf_n on a non-empty set X, if there exists a convergent series of positive terms ΣM_n such that for each $n \in \mathbb{N}$*

$$|f_n(x)| \leq M_n \quad \text{for all } x \in X,$$

then Σf_n is uniformly convergent.

Proof. Since ΣM_n is convergent we have by comparison that $\Sigma \|f_n\|_\infty$ is convergent. That is, in $(B(X), \|\cdot\|_\infty)$, Σf_n is absolutely convergent and so by Theorem 3.45 is convergent. $\square$

3.50 Example. The series of functions $\sum \dfrac{\sin nt}{n^2+t}$ for all $t \geq 0$, is uniformly convergent since

$$\left|\frac{\sin nt}{n^2+t}\right| \leq \frac{1}{n^2+t} \leq \frac{1}{n^2} \quad \text{for all } t \geq 0$$

and $\sum \dfrac{1}{n^2}$ is convergent. $\square$

We should note the following significant uniform convergence property for series of continuous functions on an interval J.

3.51 Theorem. *If a series of functions Σf_n is uniformly convergent to a function s on J and for every $n \in \mathbb{N}$, f_n is continuous at $t_0 \in J$ then s is continuous at t_0;* (which implies that the sum of a uniformly convergent series of continuous functions is continuous).

Proof. Since f_n is continuous at t_0 for each $n \in \mathbb{N}$, so $s_n \equiv \sum\limits_{k=1}^{n} f_k$ is continuous at t_0 for each $n \in \mathbb{N}$. Then the result follows from Theorem 3.14. $\square$

3.52 Remark. This theorem can be used contrapositively. For example, the Fourier series for any discontinuous wave function can never be uniformly convergent on an interval containing a discontinuity. $\square$

3.53 Exercises.

1.   Prove that the following sequences $\{x_n\}$ converge to a limit x by finding, for a given $\varepsilon > 0$, a $\nu \in \mathbb{N}$ depending on ε, such that

$$d(x_n, x) < \varepsilon \quad \text{for all } n > \nu.$$

(i)   In $\mathbb{R}$ with the usual norm, where

(a) $x_n = \frac{n-1}{2n+3}$,   (b) $x_n = \sqrt{n+1} - \sqrt{n}$.

(ii)   In $\mathbb{C}$ with the usual norm, where

(a) $x_n = \frac{1}{n} e^{in}$,   (b) $x_n = \frac{n^2+1}{n^2} + i \frac{n^2-1}{n^2}$.

(iii)   In $(\mathbb{R}^2, \|\cdot\|_2)$, where

(a) $x_n = \left(\frac{n}{n+1}, \frac{1}{n}\right)$,   (b) $x_n = \left(\frac{1+n^2}{n^2}, \frac{1-n^2}{n^2}\right)$.

(iv)   In $(C[0,1], \|\cdot\|_\infty)$, where

(a) $x_n(t) = \frac{1}{n} \sin nt$, (b) $x_n(t) = te^{-nt}$.

(v)   In $(c_0, \|\cdot\|_\infty)$, where

(a) $x_1 = \{1, 0, \ldots \qquad\}$

$x_2 = \{1, \frac{1}{2}, 0, \ldots \qquad\}$

$\cdots$

$x_n = \{1, \frac{1}{2}, \ldots, \frac{1}{2^n}, 0, \ldots\}$

$\cdots$

(b) $x_1 = \{2, \frac{3}{2}, \frac{4}{3}, \ldots \qquad\}$

$x_2 = \{\frac{3}{2}, \frac{4}{4}, \frac{5}{6}, \ldots \qquad\}$

$\cdots$

$x_n = \{\frac{n+1}{n}, \frac{n+2}{2n}, \frac{n+3}{3n}, \ldots\}$

$\cdots$

2.   Determine whether the following sequences $\{x_n\}$ converge.

(i)   In $\mathbb{R}$ with the usual norm, where

(a) $x_n = \dfrac{(-1)^n + n^2 \sin \frac{1}{n}}{n}$ (b) $x_n = n^2 \left(\frac{1}{2}\right)^n$

(ii)   In $\mathbb{C}$ with the usual norm, where

(a) $x_n = e^{-\frac{2n\pi i}{3}}$ (b) $x_n = n\left(e^{\frac{2\pi i}{n}} - 1\right)$

(iii) In $(\mathbb{R}^2, \|\cdot\|_2)$, where

(a) $x_n = \left(1 + \dfrac{(-1)^n}{n}, \; (-1)^n + \frac{1}{n}\right)$,

(b) $x_n = \left(\sqrt{\dfrac{n-1}{n}}, \; \sqrt{\dfrac{n}{n-1}}\right)$

(iv)   In $(C[0,1], \|\cdot\|_\infty)$, where

(a) $x_n(t) = nt \quad 0 \le t < \frac{1}{n}$
$\qquad\qquad \dfrac{1}{nt} \quad \frac{1}{n} < t \le 1$,

(b) $x_n(t) = \dfrac{nt^2}{1+nt}$.

(v)   In $(\ell_1, \|\cdot\|_1)$, where

(a) $x_1 \equiv \{\frac{1}{2}, 0, \ldots \qquad\qquad\}$
$\quad x_2 \equiv \{\frac{1}{2}, \frac{1}{2^2}, 0, \ldots \qquad\quad\}$
$\quad \cdots$
$\quad x_n \equiv \{\frac{1}{2}, \frac{1}{2^2}, \ldots, \frac{1}{2^n}, 0, \ldots\}$
$\quad \cdots$

(b) $x_1 = \{1, 0, \ldots \qquad\qquad\}$
$\quad x_2 = \{\frac{1}{2}, \frac{1}{2}, 0, \ldots \qquad\quad\}$
$\quad \cdots$
$\quad x_n = \{\frac{1}{n}, \frac{1}{n}, \ldots, \frac{1}{n}, 0, \ldots\}$

3. (i)   Show that the sequence $\{f_n\}$ in $C[0,1]$ where

$$f_n(t) = nte^{-nt}$$

is pointwise convergent to $\underset{\sim}{0}$ but is not uniformly convergent to $\underset{\sim}{0}$.

(ii) Prove, or disprove by counter-example, the following proposition on convergence in $(C[0,1], \|\cdot\|_1)$:

A sequence $\{f_n\}$ is convergent to f if and only if $\{f_n\}$ is pointwise convergent to f.

4. (i)   Consider $P_3[0,1]$ the linear subspace of $C[0,1]$ consisting of all cubic polynomials. Prove that a sequence $\{p_n\}$ where

$$p_n(t) = a_{n0} + a_{n1}t + a_{n2}t^2 + a_{n3}t^3$$

is convergent in $(P_3[0,1], \|\cdot\|_\infty)$ to p where

$$p(t) = a_0 + a_1 t + a_2 t^2 + a_3 t^3$$

if and only if each coefficient sequence $\{a_{nk}\}$ is convergent to a_k for each $k \in \{0,1,2,3\}$.

(ii) Consider $P[0,1]$ the linear subspace of $C[0,1]$ consisting of all polynomials.

(a) Show that the sequence $\{p_n\}$ where

$$p_n(t) = t^n$$

has the property that its coefficient sequences converge but the sequence $\{p_n\}$ does not converge in $(P[0,1], \|\cdot\|_\infty)$.

(b) By adapting the sequence $\{p_n\}$ in (a), or otherwise, show that although a sequence $\{p_n\}$ may have the property that its coefficient sequences converge, the sequence $\{p_n\}$ is not necessarily bounded.

(c) Prove that the sequence $\{p_n\}$ where

$$p_n(t) = t(1-t)^n$$

is convergent in $(P[0,1], \|\cdot\|_\infty)$ but the coefficient sequences do not converge to the coefficients of the limit polynomial.

(d) By adapting the sequence $\{p_n\}$ in (c), or otherwise, show that although a sequence $\{p_n\}$ may be convergent in $(P[0,1], \|\cdot\|_\infty)$, the coefficient sequences are not necessarily bounded.

5.
  (i) if a sequence $\{x_n\}$ is convergent to x then every subsequence $\{x_{n_k}\}$ converges to x,

  (ii) if a Cauchy sequence $\{x_n\}$ has a convergent subsequence then $\{x_n\}$ is convergent.

6. (i) For a metric space (X,d), prove that a Cauchy sequence has the property that

$$d(x_n, x_{n+1}) \to 0 \quad \text{as} \quad n \to \infty.$$

  (ii) Give an example of a sequence $\{x_n\}$ in $\mathbb{R}$ with the usual metric which has this property but which is not a Cauchy sequence.

  (iii) For a metric space (X,d) with metric d satisfying the ultrametric inequality,

$$d(x,z) \leq \max\{d(x,y), d(y,z)\} \quad \text{for all } x,y,z \in X,$$

prove that a sequence $\{x_n\}$ is Cauchy if and only if $d(x_n, x_{n+1}) \to 0$ as $n \to \infty$. (See Exercise 2.33.5)

7. (i) In a metric space (X,d) we are given that $\{x_n\}$ and $\{y_n\}$ are Cauchy sequences. Prove that $\{d(x_n, y_n)\}$ is a convergent sequence of real numbers.

  (ii) Give an example to show that it is possible to have sequences $\{x_n\}$ and $\{y_n\}$ in (X,d) such that $\{d(x_n, y_n)\}$ is a convergent sequence of real numbers but $\{x_n\}$ and $\{y_n\}$ are not Cauchy in (X,d).

8. (i)   For the linear space ℓ_1, show that no pair of the norms
           $\|\cdot\|_1, \|\cdot\|_2$ and $\|\cdot\|_\infty$ are equivalent norms for ℓ_1.

    (ii)   A linear space X has norms $\|\cdot\|$ and $\|\cdot\|'$ with the property that
           any sequence $\{x_n\}$ is Cauchy in $(X, \|\cdot\|)$ if and only if $\{x_n\}$ is
           Cauchy in $(X, \|\cdot\|')$. Prove that $\|\cdot\|$ and $\|\cdot\|'$ are equivalent
           norms for X.

9.   Consider the open interval $(-1,1)$ with the usual metric and
           with the metric d defined by

           $$d(x,y) = \left| \tan \frac{\pi x}{2} - \tan \frac{\pi y}{2} \right|.$$

    (i)   Prove that d is equivalent to the usual metric for $(-1,1)$.

    (ii)   Use this example to show that

           (a) boundedness is not an invariant under equivalent metrics,

           (b) completeness is not an invariant under equivalent metrics,
                and

           (c) the Cauchy property for sequences is not an invariant
                under equivalent metrics.

10. (i)   Show that the normed linear space $(C[0,1], \|\cdot\|_1)$ is not
           complete.
           (Hint: Approximate f by continuous functions where
           $$f(t) = 1 \quad 0 \leqslant t < \tfrac{1}{2}$$
           $$\;\;\;\;= 0 \quad \tfrac{1}{2} \leqslant t \leqslant 1 . \;)$$

    (ii)   Prove that the normed linear space $(\ell_1, \|\cdot\|_1)$ is complete.

    (iii) Determine whether the following metric subspaces are complete.
           (a) The set E of sequences containing only entries 0 and 1
                in $(m, \|\cdot\|_\infty)$.

           (b) The unit sphere in any Banach space.

11. (i)   Prove that if a sequence $\{f_n\}$ is convergent in the normed
          linear space $(C[0,1], \|\cdot\|_\infty)$ then it is convergent in
          $(C[0,1], \|\cdot\|_1)$, but give an example to show that the converse
          is not true.

    (ii) Determine whether the following series Σf_n are convergent in
          $(C[0,1], \|\cdot\|_\infty)$ and $(C[0,1], \|\cdot\|_1)$ where

          (a) $f_n(t) = \dfrac{t^n}{n!}$ (b) $f_n(t) = \dfrac{t^n}{n}$.

    (iii) Deduce that neither $(P[0,1], \|\cdot\|_\infty)$ nor $(P[0,1], \|\cdot\|_1)$ is
          complete.

12. (i)   The function $f(t) = |t|$ on $[-\pi, \pi]$ has Fourier cosine series

          $$\frac{\pi}{2} - \frac{4}{\pi}\left(\frac{\cos t}{1^2} + \frac{\cos 3t}{3^2} + \ldots + \frac{\cos(2n-1)t}{(2n-1)^2} + \ldots\right) .$$

          Determine whether the series is uniformly convergent on
          $[-\pi, \pi]$.

    (ii) The function $f(t) = t$ on $[-\pi, \pi]$ has Fourier sine series

          $$2\left(\sin t - \frac{\sin 2t}{2} + \ldots + (-1)^{n-1}\frac{\sin nt}{n} + \ldots\right) .$$

          Determine whether the series is uniformly convergent on $[-\pi, \pi]$.

13. (i)   The exponential function $f(t) = e^t$ has Taylor series at $t = 0$,

          $$1 + t + \frac{t^2}{2!} + \ldots + \frac{t^n}{n!} + \ldots$$

          (a) Prove that the series converges pointwise to the exponential
               function on $\mathbb{R}$.

          (b) Prove that the series is uniformly convergent on any
               interval $[-a, a]$ where $a > 0$.

          (c) But show that the series is not uniformly convergent on $\mathbb{R}$.

13. (ii)   The logarithmic function $f(t) = \ln(1+t)$ has Taylor series at $t = 0$,

$$t - \frac{t^2}{2} + \frac{t^3}{3} + \ldots + (-1)^{n+1} \frac{t^n}{n} + \ldots$$

(a) Prove that the series converges pointwise to the generating logarithmic function on $(-1,1]$.

(b) Prove that the series is uniformly convergent on any interval $[-a,1]$ where $0 < a < 1$.

(c) But show that the series is not uniformly convergent on $(-1,1]$.

14.   A normed linear space $(X, \|\cdot\|)$ is said to have a *Schauder basis* if there exists a sequence $\{e_1, e_2, \ldots, e_n, \ldots\}$ in X where $\|e_n\| = 1$ for all $n \in \mathbb{N}$ and for every $x \in X$ there exists a unique sequence of scalars $\{\lambda_1, \lambda_2, \ldots, \lambda_n, \ldots\}$ such that the series $\Sigma \lambda_n e_n$ is convergent to x. Prove that such an X is finite dimensional if it has the property that every sequence $\{x_n\}$ where

$$x_1 \equiv \lambda_{11} e_1 + \lambda_{12} e_2 + \ldots + \lambda_{1n} e_n + \ldots$$
$$x_2 \equiv \lambda_{21} e_1 + \lambda_{22} e_2 + \ldots + \lambda_{2n} e_n + \ldots$$
$$\cdot \quad \cdot \quad \cdot$$
$$x_n \equiv \lambda_{n1} e_1 + \lambda_{n2} e_2 + \ldots + \lambda_{nn} e_n + \ldots$$
$$\cdot \quad \cdot \quad \cdot$$

is convergent to x where

$$x \equiv \lambda_1 e_1 + \lambda_2 e_2 + \ldots + \lambda_n e_n + \ldots$$

if and only if each co-ordinate sequence $\{\lambda_{nk}\}$ is convergent to λ_k for each $k \in \mathbb{N}$.

(Hint: Show that in an infinite dimensional normed linear space $(X, \|\cdot\|)$ with a Schauder basis it is possible to find a sequence $\{x_n\}$ where each co-ordinate sequence $\{\lambda_{nk}\}$ is convergent to 0 but $\{x_n\}$ is not convergent to $\underset{\sim}{0}$.)

15.   Given an incomplete normed linear space $(X, \|\cdot\|)$ it is possible
to construct a complete normed linear space $(\tilde{X}, \|\cdot\|)$, called the *completion*
of $(X, \|\cdot\|)$, closely associated with the original space $(X, \|\cdot\|)$ in a
natural way:

Given $(X, \|\cdot\|)$, denote by $X^\dagger$ the set of Cauchy sequences in $(X, \|\cdot\|)$.

    (i)   Prove that $X^\dagger$ is a linear space under pointwise definition of
the linear space operations and that the norm $\|\cdot\|$ on X
generates a semi-norm $p^\dagger$ on $X^\dagger$ defined for $x^\dagger \equiv \{x_n\} \in X^\dagger$
by $p^\dagger(x^\dagger) = \lim \|x_n\|$.

    (ii)   Prove that the normed linear space $\left(\dfrac{X^\dagger}{\ker p^\dagger}, \|\cdot\|^\dagger\right)$ associated
with $(X^\dagger, p^\dagger)$ defined as in Theorem 1.32 is complete.

(It is the Banach space $\left(\dfrac{X^\dagger}{\ker p^\dagger}, \|\cdot\|^\dagger\right)$ which is called the completion of
$(X, \|\cdot\|)$ and is briefly denoted $(\tilde{X}, \|\cdot\|)$.)

4. CLUSTER POINTS AND CLOSURE

We now introduce a more general limit process associated with any set but which is nevertheless characterised by the convergence of sequences selected from the set.

4.1 Definition.   Given a set A in a metric space (X,d), an element $x \in X$ is said to be a *cluster point* of A if there exists a sequence $\{a_n\}$ in $A\backslash\{x\}$ which is convergent to X.

4.2 Examples.   Consider $\mathbb{R}$ with the usual norm.

(i)   Set set $A \equiv \{(-1)^n + \frac{1}{n} : n \in \mathbb{N}\}$ has -1 and +1 as cluster points because the sequence $\{-1 + \frac{1}{2n-1}\}$ in A is convergent to -1 and the sequence $\{1 + \frac{1}{2n}\}$ in A is convergent to +1.

(ii)   The set $\mathbb{Q}$ of rationals has all the points of $\mathbb{R}$ as cluster points:
For any rational r, the sequence $\{r + \frac{1}{n}\}$ in $\mathbb{Q}$ is convergent to r and for any irrational s, the density property of $\mathbb{Q}$ in $\mathbb{R}$ implies that there exists a sequence of rational approximations to s, $\{r_n\}$ convergent to s. □

4.3 Examples.   Consider $(\mathbb{R}^2, \|\cdot\|_2)$.

(i)   The set $A \equiv \{((-1)^n \frac{1}{n}, (-1)^n) : n \in \mathbb{N}\}$ has $(0,1)$ and $(0,-1)$ as cluster points because the sequence $\{(\frac{1}{2n}, 1)\}$ is convergent to $(0,1)$ and the sequence $\{(\frac{1}{2n-1}, -1)\}$ is convergent to $(0,-1)$.

(ii)   The set $A \equiv \{(\frac{p}{q}, \frac{1}{q}) : p,q \text{ mutually prime}, p,q \in \mathbb{N}\}$ has all points of $\{(\lambda, 0) : \lambda \geqslant 0\}$ as cluster points:
Given any $\lambda \geqslant 0$ there exists a sequence of rational approximations to λ, $\{\frac{p_n}{q_n}\}$ in $\mathbb{R}^+\backslash\{\lambda\}$, convergent to λ. Now given $\varepsilon > 0$, there exists only a finite number of rationals of the form $\frac{1}{q}$ which are greater than ε. So $\{\frac{1}{q_n}\}$ is convergent to 0. It follows from Example 3.7 that $\{(\frac{p_n}{q_n}, \frac{1}{q_n})\}$ is convergent to $(\lambda, 0)$. □

4.4 Example.   Consider $(C[0,1], \|\cdot\|_\infty)$ and the linear subspace of polygonal functions L on $[0,1]$. For any partition $P \equiv \{t_0, t_1, \ldots, t_n\}$ of $[0,1]$ set of real numbers $\{\alpha_0, \alpha_1, \ldots, \alpha_n\}$ a polygonal function g on $[0,1]$ is defined for each $k \in \{0, 1, \ldots, n\}$ by

$$g(t_k) = \alpha_k$$

and on each subinterval $[t_k, t_{k+1}]$ by

$$g(\lambda t_k + (1-\lambda) t_{k+1})$$
$$= \lambda \alpha_k + (1-\lambda) \alpha_{k+1} \quad \text{for all } 0 \leqslant \lambda \leqslant 1.$$

We show that any continuous function f on $[0,1]$ is a cluster point of the set L:

        Suppose f is not polygonal. Since f is uniformly continuous on $[0,1]$, for every $n \in \mathbb{N}$ there exists a $\delta_n > 0$ such that

$$\left| f(s) - f(t) \right| < \frac{1}{n} \quad \text{for } \left| s - t \right| < \delta_n.$$

Choose a partition $P \equiv \{t_0, t_1, \ldots, t_n\}$ of $[0,1]$ such that $\left| t_{k+1} - t_k \right| < \delta_n$ for each $k \in \{0,1,\ldots,n-1\}$ and define a polygonal function g_n on $[0,1]$ based on this partition P and taking $\alpha_k = f(t_k)$ for each $k \in \{0,1,\ldots,n\}$. Then $\left| f(t) - g_n(t) \right| < \frac{1}{n}$ for all $t \in [0,1]$ and so $\| f - g_n \|_\infty < \frac{1}{n}$; that is, the sequence $\{g_n\}$ is convergent to f.

        When f is polygonal based on a partition $P \equiv \{t_0, t_1, \ldots, t_n\}$ we can construct an approximating sequence of polygonal functions $\{g_n\}$ based on the same partition P but with function values

$$g_n(t_0) = f(t_0) + \frac{1}{n}$$
$$g_n(t_k) = f(t_k) \quad \text{for each } k \in \{1,2,\ldots,n\}. \quad \square$$

4.5 Remark. Whether or not a point is a cluster point of a set depends on the metric defined on the space. For example, in a discrete metric space (X,d), the subsets have no cluster points: Given a subset A and $x \in X$, $d(x,a) = 1$ for all $a \in A \backslash \{x\}$, so no sequence in $A \backslash \{x\}$ can converge to x. $\square$

4.6 Remark. However, it is clear from the Definitions 4.1 and 3.18 that the property of a point being a cluster point of a set is invariant under equivalent metrics. $\square$

The following alternative characterisation of cluster points is useful.

4.7 Theorem. *Given a set* A *in a metric space* (X,d) *a point* $x \in X$ *is a cluster point of* A *if and only if for any given* $\varepsilon > 0$,

$$\left(B(x;\varepsilon)\backslash\{x\}\right) \cap A \neq \phi.$$

Proof. Suppose that x is a cluster point of A; that is, there exists a sequence $\{a_n\}$ in $A\backslash\{x\}$ such that $\{a_n\}$ is convergent to x. Then for any given $\varepsilon > 0$ there exists a $\nu \in \mathbb{N}$ such that

$$a_n \in B(x;\varepsilon)\backslash\{x\} \quad \text{for all } n > \nu.$$

But then for any given $\varepsilon > 0$,

$$\left(B(x;\varepsilon)\backslash\{x\}\right) \cap A \neq \phi.$$

Conversely, suppose that for any given $\varepsilon > 0$

$$\left(B(x;\varepsilon)\backslash\{x\}\right) \cap A \neq \phi.$$

Then there exists an $a_1 \in A$ such that $a_1 \in B(x;1)\backslash\{x\}$. Write $r_1 \equiv \min\{d(x,a_1);\frac{1}{2}\}$. Then there exists an $a_2 \in A$ such that $a_2 \in B(x;r_1)\backslash\{x\}$.
We inductively define a sequence $\{a_n\}$ in $A\backslash\{x\}$ such that

$$d(x,a_n) < \frac{1}{2^{n-1}} \quad \text{for all } n \in \mathbb{N};$$

that is, $\{a_n\}$ is convergent to x. $\square$

It is clear from the examples that not every member of a set is a cluster point of the set.

4.8 Definition. Given a set A in a metric space (X,d), a point $a \in A$ which is not a cluster point of A is said to be an *isolated point* of A.

4.9 Remark. It is clear from Theorem 4.7 that a $\in$ A is an isolated point
of A if and only if there exists an r > 0 such that

$$B(a;r) \cap A = \{a\}. \quad \square$$

4.10 Example. In Example 4.3(ii) all the points of A are isolated points:
Consider any point $\left(\frac{p_0}{q_0}, \frac{1}{q_0}\right)$ in A. Since there exists only a finite number
of rationals of the form $\frac{1}{q}$ which are greater than $\frac{1}{2q_0}$, so there exists a
point of A other than $\left(\frac{p_0}{q_0}, \frac{1}{q_0}\right)$, say $\left(\frac{p'}{q'}, \frac{1}{q'}\right)$ which is closest to $\left(\frac{p_0}{q_0}, \frac{1}{q_0}\right)$.
Writing $\left\| \left(\frac{p_0}{q_0}, \frac{1}{q_0}\right) - \left(\frac{p'}{q'}, \frac{1}{q'}\right) \right\|_2 \equiv r$ then $B\left(\left(\frac{p_0}{q_0}, \frac{1}{q_0}\right);r\right) \cap A = \left\{\left(\frac{p_0}{q_0}, \frac{1}{q_0}\right)\right\}. \quad \square$

The examples show that a cluster point of a set may or may not
be a member of the set.

4.11 Definitions. Given a set A in a metric space (X,d), the set of
cluster points of A is called the *derived set* of A and is denoted by A'.

The set A is said to be *closed* if it contains all its cluster points;
that is, if A' $\subseteq$ A.

4.12 Examples.
  (i)   It is clear that every finite subset of any metric space (X,d)
          is closed.
  (ii)   From Remark 4.5, we deduce that all subsets of a discrete
          metric space (X,d) are closed.

In Definition 2.1 we introduced the idea of a closed ball in a
metric space. We should show that the terminology we have introduced in
Definition 4.11 is consistent with that of Definition 2.1.

4.13 Theorem. *In any metric space* (X,d), *every closed ball is a closed
set.*

Proof. Given a $\in$ X and r > 0 consider the closed ball B[a;r]. Consider
any cluster point x of B[a;r]. Now there exists a sequence $\{a_n\}$ in
B[a;r]\{x} which is convergent to x. But

$$d(a,x) \leq d(a,a_n) + d(a_n,x)$$

and because $d(a_n,x) \to 0$ as $n \to \infty$ we deduce that $d(a,x) \leqslant r$. So
$x \in B[a;r]$ and we conclude that $B[a;r]$ is a closed set. ☐

4.14 Remark. This theorem also illustrates a technique for proving sets
closed: We consider a cluster point, take an approximating sequence in
the set and use the particular properties of the set to conclude that the
cluster point belongs to the set. ☐

The following characterisation of closed sets is an immediate
deduction from Theorem 4.7.

4.15 Theorem. *A set A in a metric space* (X,d) *is closed if and only if
for every* $x \in C(A)$ *there exists an* $r > 0$ *such that*

$$B(x;r) \cap A = \phi.$$

It is often convenient to use this result to prove that a set
is closed.

4.16 Example. The subset E of the closed interval $[0,1]$ consisting of
those real numbers which have a decimal representation of the form

$$\sum_{n=1}^{\infty} \frac{a_n}{10^n} \quad \text{where } a_n \in \{0,1\} \text{ for all } n \in \mathbb{N}$$

is a closed subset of $[0,1]$ with the usual metric:
Now $\cdot01$ and $\cdot0\dot{9} \equiv \cdot1$ are members of E. Consider any $y \in [0,1] \backslash E$. There
exists a first digit in the decimal representation of y which is not 0 or
1. Suppose it is the n_0th digit. Then $d(y,E) > \dfrac{1}{10^{n_0+1}}$,
so $B\left(y; \dfrac{1}{10^{n_0+1}}\right) \cap E = \phi$ and we conclude that E is closed. ☐

4.17 Remark. It follows from Definition 4.11 and Remark 4.6 that the
property of a set being closed is invariant under equivalent metrics. ☐

In any metric space, the family of closed sets has the
following fundamental properties.

<u>4.18 Theorem</u>. *Given a metric space* (X,d),

    (i)   *both ϕ and X are closed,*

    (ii)   *any intersection of closed sets is closed, and*

    (iii)   *any union of a finite number of closed sets is closed.*

<u>Proof</u>. (i) is immediate from Definition 4.11.

    (ii) Consider a family $\{F_\alpha\}$ of closed sets and x a cluster point of $\cap F_\alpha$. Now there exists a sequence $\{x_n\}$ in $\cap F_\alpha \backslash \{x\}$ such that $\{x_n\}$ is convergent to x. But then, for each α, $\{x_n\}$ lies in $F_\alpha \backslash \{x\}$ so x is a cluster point of F_α. But F_α is closed so $x \in F_\alpha$. Therefore $x \in \cap F_\alpha$ and we conclude that $\cap F_\alpha$ is closed.

    (iii) Consider a finite family $\{F_k : k \in \{1,2,\ldots,n\}\}$ of closed sets and $x \in C\left(\bigcup_1^n F_k\right)$. Then for each $k \in \{1,2,\ldots,n\}$, $x \in C(F_k)$. Since F_k is closed we have by Theorem 4.15 that there exists an $r_k > 0$ such that

$$B(x;r_k) \cap F_k = \phi.$$

Writing $r \equiv \min\{r_k : k \in \{1,2,\ldots,n\}\}$ we have that

$$B(x;r) \cap \bigcup_1^n F_k = \phi.$$

So again by Theorem 4.15 we conclude that $\bigcup_1^n F_k$ is closed. $\square$

<u>4.19 Remark</u>. We note that in Theorem 4.18(iii) the finiteness condition is important because a union of an infinite family of closed sets need not necessarily be closed:

In $\mathbb{R}$ with the usual norm, consider the family of closed intervals $\left\{ \left[\frac{1-k}{k}, \frac{k-1}{k} \right] : k \in \mathbb{N} \right\}$. Now $\bigcup_1^\infty \left\{ \left[\frac{1-k}{k}, \frac{k-1}{k} \right] \right\} = (-1,1)$

an open interval which is not a closed set. $\square$

    We now establish an important link between closedness and completeness. The significance of this link follows from the fact that whether or not a set is closed is relative to the space in which it is a subset, but completeness is not a relative concept.

4.20 Theorem. *Given a metric space* (X,d) *and a subset* Y *of* X,

    (i)   *if* $(Y,d|_Y)$ *is complete then* Y *is closed in* (X,d),

    (ii)   *if* (X,d) *is complete and* Y *is closed in* (X,d) *then* $(Y,d|_Y)$

           is complete, ((ii) says that a closed subset of a complete

           space is complete).

Proof.   (i) Consider x a cluster point of Y in (X,d). Then there exists a sequence $\{y_n\}$ in Y\{x} such that $\{y_n\}$ is convergent to x. Then $\{y_n\}$ is Cauchy in $(Y,d|_Y)$. But we are given that $(Y,d|_Y)$ is complete so x $\in$ Y and we conclude that Y is closed.

        (ii) Consider $\{y_n\}$ a Cauchy sequence in $(Y,d|_Y)$. Then $\{y_n\}$ is Cauchy in (X,d). But (X,d) is complete so there exists an x $\in$ X such that $\{y_n\}$ is convergent to x. We deduce that either x $\in$ Y or x is a cluster point of Y. But we are given that Y is closed in (X,d) so x $\in$ Y and we conclude that $(Y,d|_Y)$ is complete. $\square$

    This theorem has an interesting implication for finite dimensional spaces.

4.21 Corollary. *In a normed linear space* $(X, \|\cdot\|)$, *any finite dimensional linear subspace* Y_m *is closed.*

Proof. From Theorem 3.36, the finite dimensional normed linear space $(Y_m, \|\cdot\|_{Y_m})$ is complete, so the proof is immediate from Theorem 4.20(i). $\square$

    Theorem 4.20(ii) also provides a useful method for establishing the completeness of certain spaces which can readily be identified as subspaces of complete normed linear spaces.

    The following examples are significant subspaces of the Banach space $(B(J), \|\cdot\|_\infty)$ where J is an interval.

4.22 Example. *For* $C(J)$ *the linear space of bounded continuous functions on* J, *(which includes as a special case* $C[a,b]$ *the linear space of continuous functions on* [a,b]), $(C(J), \|\cdot\|_\infty)$ *is complete.*

Proof. From Theorem 4.20(ii) it is sufficient to show that $C(J)$ is a closed subset of $(B(J), \|\cdot\|_\infty)$. Consider f a cluster point of $C(J)$ in

$(B(J), \|\cdot\|_\infty)$. Then there exists a sequence $\{f_n\}$ in $C(J)\backslash\{f\}$ such that $\{f_n\}$ is uniformly convergent to f. But then from Theorem 3.14 it follows that f is continuous on J. So $f \in C(J)$ and $C(J)$ is closed. $\square$

4.23 Example. *The normed linear space* $(R[a,b], \|\cdot\|_\infty)$, *of Riemann integrable functions on* $[a,b]$ *with the supremum norm, is complete.*

Proof. Again from Theorem 4.20(ii) it is sufficient to show that $R[a,b]$ is a closed subset of $(B[a,b], \|\cdot\|_\infty)$. Consider f a cluster point of $R[a,b]$ in $(B[a,b], \|\cdot\|_\infty)$. Then there exists a sequence $\{f_n\}$ in $R[a,b]\backslash\{f\}$ such that $\{f_n\}$ is uniformly convergent to f; that is, given $\varepsilon > 0$ there exists a $\nu \in \mathbb{N}$ such that

$$\|f_n - f\|_\infty < \varepsilon \quad \text{for all } n > \nu.$$

So for every $t \in [a,b]$,

$$|f_n(t) - f(t)| < \varepsilon \quad \text{for all } n > \nu.$$

Since each f_n is Riemann integrable on $[a,b]$, $f_{\nu+1}$ satisfies a Riemann condition; that is, there exists a partition P of $[a,b]$ such that

$$U(P, f_{\nu+1}) - L(P, f_{\nu+1}) < \varepsilon.$$

Therefore, for this partition P of $[a,b]$,

$$U(P,f) < U(P, f_{\nu+1}) + \varepsilon(b-a) \quad \text{and}$$
$$L(P,f) > L(P, f_{\nu+1}) - \varepsilon(b-a).$$

It follows that

$$U(P,f) - L(P,f) < U(P, f_{\nu+1}) - L(P, f_{\nu+1}) + 2\varepsilon(b-a)$$
$$< \varepsilon\big(1 + 2(b-a)\big);$$

that is, f satisfies a Riemann condition so f is Riemann integrable on $[a,b]$. $\square$

Theorem 4.20(i) can be used to show that certain normed linear spaces are incomplete. The next example is of a highly structured subspace of $(B[a,b], \|\cdot\|_\infty)$ which is nevertheless not complete.

4.24 Example. *The normed linear space* $(C^1[a,b], \|\cdot\|_\infty)$ *of continuously differentiable functions on* $[a,b]$ *with the supremum norm, is not complete.*

Proof. Consider the function $f \in C[-1,1]$ defined by

$$f(t) = |t|.$$

Now f is not differentiable at 0. Nevertheless f is a cluster point of $C^1[-1,1]$ in $(C[-1,1], \|\cdot\|_\infty)$ because the sequence $\{f_n\}$ in $C^1[-1,1]$ where

$$f_n(t) = \sqrt{t^2 + \frac{1}{n^2}}$$

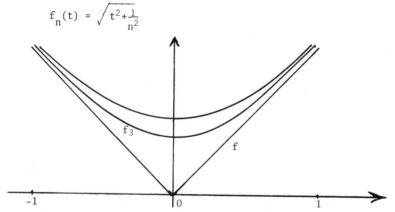

Figure 8.   The functions $f_n(t) = \sqrt{t^2 + \frac{1}{n}}$ converging uniformly to the function $f(t) = |t|$.

satisfies $\|f_n - f\|_\infty = \frac{1}{n}$ and so is uniformly convergent to f. Therefore $C^1[-1,1]$ is not closed in $(C[-1,1], \|\cdot\|_\infty)$ and by Theorem 4.20(i), $(C^1[-1,1], \|\cdot\|_\infty)$ is not complete. $\square$

In a metric space it is useful to associate with any subset a related closed set.

4.25 Definition. Given a metric space (X,d), the *closure* of a subset A is the set $A \cup A'$ which is denoted by $\overline{A}$.

In terms of our Definition 4.1 for cluster points and our characterisation Theorem 4.7 for cluster points we can give the following characterisations of closure.

4.26 Theorem. *Given a subset* A *of a metric space* (X,d),

    (i) $x \in \overline{A}$ *if and only if there exists a sequence* $\{a_n\}$ *in* A *such that* $\{a_n\}$ *is convergent to* x,

    (ii) $x \in \overline{A}$ *if and only if for any given* $\varepsilon > 0$, $B(x;\varepsilon) \cap A \neq \phi$.

4.27 Remark. In terms of this notion we can say that a subset A of a metric space (X,d) is closed if and only if $\overline{A} = A$. $\square$

However, we need to determine general properties for the closure of a set.

4.28 Theorem. *Given a subset* A *of a metric space* (X,d),

    (i) $\overline{A}$ *is closed,*

    (ii) $\overline{A}$ *is the smallest closed set containing* A.

Proof.   (i)   Consider x a cluster point of $\overline{A}$. Then there exists a sequence $\{\overline{a}_n\}$ in $\overline{A}\backslash\{x\}$ such that $\{\overline{a}_n\}$ is convergent to x. We modify the sequence $\{\overline{a}_n\}$ to obtain a sequence $\{a_n\}$ in A such that $\{a_n\}$ is convergent to x:

For each $n \in \mathbb{N}$, $\overline{a}_n$ is a member of A or is a cluster point of A. If $\overline{a}_n$ is a member of A, write $a_n \equiv \overline{a}_n$. If $\overline{a}_n$ is a cluster point of A and not a member of A, we have from Theorem 4.7 that we can choose $a_n \in A\backslash\{\overline{a}_n\}$ such that $d(a_n, \overline{a}_n) < \frac{1}{n}$. Then the sequence $\{a_n\}$ is convergent to x. From Theorem 4.26(i) we deduce that $x \in \overline{A}$ and we conclude that $\overline{A}$ is closed.

    (ii)   Consider any closed set F such that $A \subseteq F$. We show that there are no cluster points of A in $C(F)$:

Consider any $x \in C(F)$. Since F is closed we deduce from Theorem 4.15 that there exists an $r > 0$ such that $B(x;r) \cap F = \phi$. Then $B(x;r) \cap A = \phi$ and from Theorem 4.7 we have that x is not a cluster point of A. So $A' \subseteq F$ and $\overline{A} \subseteq F$. $\square$

4.29 Example. *In* $(m, \|\cdot\|_\infty)$, $\overline{E}_0 = c_0$.

Proof.   Consider $x \equiv \{\lambda_1, \lambda_2, \ldots, \lambda_n, \ldots\} \in c_0\backslash E_0$. The sequence of truncations $\{x_n\}$ where

$$x_1 \equiv \{\lambda_1, 0, \dots \qquad \qquad \}$$

$$x_2 \equiv \{\lambda_1, \lambda_2, 0, \dots \qquad \}$$

. . .

$$x_n \equiv \{\lambda_1, \lambda_2, \dots, \lambda_n, 0, \dots\}$$

. . .

lies in E_0 and

$$\|x - x_n\|_\infty = \sup\{|\lambda_k| : k > n+1\}$$
$$\to 0 \quad \text{as} \quad n \to \infty$$

since $x \in c_0$. So $\overline{E}_0 \supseteq c_0$.

For $x \equiv \{\lambda_1, \lambda_2, \dots, \lambda_n, \dots\} \in m \backslash c_0$, $\lambda_n \not\to 0$ so there exists an $r > 0$ and a subsequence $\{\lambda_{n_k}\}$ such that $|\lambda_{n_k}| > r$ for all $k \in \mathbb{N}$. But then $\|x - y\|_\infty > r$ for all $y \in E_0$, so x is not a cluster point of E_0. Therefore, $\overline{E}_0 = c_0$. □

<u>4.30 Remark</u>. As with Remarks 4.6 and 4.17 on cluster points and closedness, so the closure of a set is invariant under equivalent metrics. □

We should be aware that in some metric spaces we have situations which do not accord with our intuition.

<u>4.31 Example</u>. In a discrete metric space (X, d), Example 4.12(ii) implies that every subset is its own closure. This produces some strange results: For any given $x \in X$,

$$\overline{B(x;1)} = B(x;1) = \{x\}.$$

But $B[x;1] = X$ and so

$$\overline{B(x;1)} \neq B[x;1]. \quad □$$

However, in normed linear spaces we do have a more acceptable state of affairs.

4.32 Theorem. *In a normed linear space* $(X, \|\cdot\|)$,

$$\overline{B(\underline{0};1)} = B[\underline{0};1].$$

Proof. Consider any $x \in X$, $\|x\| = 1$. For any sequence of real numbers $\{\lambda_n\}$ converging to 1 where $0 < \lambda_n < 1$, we have that $\lambda_n x \in B(\underline{0};1)$ for all $n \in \mathbb{N}$ and $\{\lambda_n x\}$ converges to x. Therefore, x is a cluster point of $B(\underline{0};1)$ and so

$$B(\underline{0};1) \subseteq B[\underline{0};1] \subseteq \overline{B(\underline{0};1)}.$$

But by Theorem 4.13, $B[\underline{0};1]$ is closed so by Theorem 4.28(ii)

$$\overline{B(\underline{0};1)} = B[\underline{0};1]. \quad \square$$

We have another significant closure property for normed linear spaces.

4.33 Theorem. *In a normed linear space* $(X, \|\cdot\|)$, *the closure* $\overline{M}$ *of a linear subspace* M *is also a linear subspace.*

Proof. Consider $\overline{x}, \overline{y} \in \overline{M}$. From Theorem 4.26(i) there exist sequences $\{x_n\}$ and $\{y_n\}$ in M such that $\{x_n\}$ converges to $\overline{x}$ and $\{y_n\}$ converges to $\overline{y}$. From Theorem 3.27 we see that sequence $\{x_n + y_n\}$ converges to $\overline{x} + \overline{y}$ and sequence $\{\alpha x_n\}$ converges to $\alpha \overline{x}$ for any scalar α. Again from Theorem 4.26(i) we deduce that $\overline{x} + \overline{y} \in \overline{M}$ and $\alpha \overline{x} \in \overline{M}$ which implies that $\overline{M}$ is a linear subspace. $\square$

We introduce another important metric space property defined in terms of closure.

4.34 Definition. Given a metric space (X,d), a subset A is said to be *dense* in X if $\overline{A} = X$.
This means that A is dense in X if and only if every point of X is either a point of A or a cluster point of A.

<u>4.35 Remark</u>. Using the form given in Theorem 4.26(ii) we have that A is dense in X if and only if for every x ∈ X and every ε > 0

$$B(x;\varepsilon) \cap A \neq \phi. \ \square$$

<u>4.36 Example</u>. In $\mathbb{R}$ with the usual norm, $\overline{\mathbb{Q}} = \mathbb{R}$, the rationals are dense, in the reals, and $\overline{\mathbb{R}\backslash\mathbb{Q}} = \mathbb{R}$, the irrationals are dense in the reals. $\square$

<u>4.37 Example</u>. *In* $(\mathbb{R}^m, \|\cdot\|_2)$, *the set* $\mathbb{Q}^m$ *is dense in* $\mathbb{R}^m$.

<u>Proof</u>. Consider any point $x \equiv (\lambda_1, \lambda_2, \ldots, \lambda_m) \in \mathbb{R}^m$. Now from Example 4.36 there exist sequences of rationals

$$\{\lambda_{1n}\} \text{ convergent to } \lambda_1,$$
$$\{\lambda_{2n}\} \text{ convergent to } \lambda_2,$$
$$\cdots$$
$$\{\lambda_{mn}\} \text{ convergent to } \lambda_m.$$

So from Example 3.7 the sequence $\{x_n\}$ in $\mathbb{Q}^m$ where

$$x_n \equiv (\lambda_{1n}, \lambda_{2n}, \ldots, \lambda_{mn})$$

is convergent to x and so $x \in \overline{\mathbb{Q}^m}$. $\square$

<u>4.38 Example</u>. We see from Example 4.29 that E_0 is dense in $(c_0, \|\cdot\|_\infty)$.

<u>4.39 Example</u>. We see from Example 4.4 that L the linear space of polygonal functions on $[0,1]$ is dense in $(C[0,1], \|\cdot\|_\infty)$. $\square$

There is a significant classification of metric spaces according to what is called density character.

<u>4.40 Definition</u>. A metric space (X,d) is said to be *separable* if it contains a countable dense subset.

4.41 Remark. In Examples 4.36 and 4.37 we see that $\mathbb{R}$ with the usual norm is separable since $\mathbb{Q}$ is countable and dense in $\mathbb{R}$ and $(\mathbb{R}^m, \|\cdot\|_2)$ is separable since $\mathbb{Q}^m$ is countable and dense in $\mathbb{R}^m$. $\square$

Separability is of significance in that it is a property possessed by every finite dimensional normed linear space.

4.42 Theorem. *A finite dimensional normed linear space* $(X_m, \|\cdot\|)$ *is separable.*

Proof. Suppose that X_m is a real linear space and that $\{e_1, e_2, \ldots, e_m\}$ is a basis for X_m where $\|e_k\| = 1$ for all $k \in \{1, 2, \ldots, m\}$. Consider $x \equiv \lambda_1 e_1 + \lambda_2 e_2 + \ldots + \lambda_m e_m$. Since $\mathbb{Q}$ is dense in $\mathbb{R}$ then, as in Example 4.37, for each λ_k, $k \in \{1, 2, \ldots, m\}$ there exist sequences of rationals

$$\{\lambda_{1n}\} \text{ convergent to } \lambda_1,$$
$$\{\lambda_{2n}\} \text{ convergent to } \lambda_2,$$
$$\ldots$$
$$\{\lambda_{mn}\} \text{ convergent to } \lambda_m.$$

From Theorem 3.8, the sequence $\{x_n\}$ where

$$x_n \equiv \lambda_{1n} e_1 + \lambda_{2n} e_2 + \ldots + \lambda_{mn} e_m$$

is convergent to x. So the set $X_{m\mathbb{Q}}$ of elements with rational coefficients is dense in $(X_m, \|\cdot\|)$. But since $\mathbb{Q}^m$ is countable then $X_{m\mathbb{Q}}$ is countable. The case when X_m is a complex linear space can be dealt with by considering it as a real linear space of $2m$ dimensions. $\square$

There are many infinite dimensional normed linear spaces which are separable.

4.43 Example. *The Banach space* $(c_0, \|\cdot\|_\infty)$ *is separable.*

Proof. Consider the real linear subspace E_0 and the set $E_{0\mathbb{Q}}$ of sequences of rationals with only a finite number of non-zero terms. Now $E_{0\mathbb{Q}}$ is countable. For any given element $x \equiv \{\lambda_1, \lambda_2, \ldots, \lambda_m, 0, \ldots\} \in E_0$ we have from Theorem 4.42 that there exists a sequence $\{x_n\}$ in $E_{0\mathbb{Q}}$ where

$$x_n \equiv \{\lambda_{1n}, \lambda_{2n}, \ldots, \lambda_{mn}, 0, \ldots\}$$

which is convergent to x.

The case when E_0 is a complex linear space can be dealt with by considering each element $x \equiv \{\lambda_1, \lambda_2, \ldots, \lambda_m, 0, \ldots\}$ as an element of a 2m-dimensional real linear subspace.

In either case $E_{0\mathbb{Q}}$ is dense in E_0. But from Example 4.29 we see that E_0 is dense in c_0. It is then straightforward to see that $E_{0\mathbb{Q}}$ is dense in c_0; (see Exercise 4.51.13(i)). □

However there are many non-separable normed linear spaces.

4.44 Example. *The Banach space* $(m, \|\cdot\|_\infty)$ *is not separable.*

Proof. Consider the set E of sequences consisting of digits 0 and 1. Now E is uncountable. Moreover, for any $x, y \in E$, $x \neq y$, $\|x-y\|_\infty = 1$. So $\{B(x; \frac{1}{2}) : x \in E\}$ is an uncountable family of disjoint open balls. For any subset S dense in m, we have from Remark 4.35 that each ball in this family must contain at least one element of S. But this implies that S is uncountable. □

We introduce a concept defined in terms of closure.

4.45 Definition. Given a metric space (X,d) and a subset A, the set $\overline{A} \cap \overline{C(A)}$ is called the *boundary* of the set A and is denoted by ∂A.

4.46 Remark. It follows from Theorem 4.18(ii) that for any set A the boundary ∂A is a closed set. □

The boundary of a set is not always the subset which its name might suggest even in familiar spaces.

4.47 Example. In $\mathbb{R}$ with the usual norm

$$\partial\mathbb{Q} = \overline{\mathbb{Q}} \cap \overline{\mathbb{R}\backslash\mathbb{Q}} = \mathbb{R} \cap \mathbb{R} = \mathbb{R}. \quad \square$$

4.48 Example. In a discrete metric space (X,d) the boundary of any subset is the null set:

From Example 4.12(ii) we have that every subset is closed. So for any subset A, $\overline{A}$ = A and $\overline{C(A)}$ = C(A) and therefore ∂A = A $\cap$ C(A) = ϕ. $\square$

However, in normed linear spaces the boundary of balls does accord with our intuition.

4.49 Theorem. *In a normed linear space* $(X, \|\cdot\|)$,

$$\partial B(\underline{0};1) = S(\underline{0};1).$$

Proof. From Theorem 4.32 we have that

$$\overline{B(\underline{0};1)} = B[\underline{0};1].$$

Consider $x \notin C\big(B(\underline{0};1)\big)$ where $\|x\| = r < 1$. For any sequence $\{x_n\}$ in $C\big(B(\underline{0};1)\big)$ we have $\|x_n\| \geq 1$ for all $n \in \mathbb{N}$ so

$$1 - r \leq \big| \|x\| - \|x_n\| \big| \leq \|x - x_n\|$$

and $\{x_n\}$ cannot converge to x. Therefore $C\big(B(\underline{0};1)\big)$ contains all its cluster points and is closed. Consequently

$$\overline{C(B(\underline{0};1))} = C\big(B(\underline{0};1)\big)$$

and $\qquad \partial B(\underline{0};1) \quad = \quad B[\underline{0};1] \cap C\big(B(\underline{0};1)\big)$

$$= \quad S(\underline{0};1) \ . \ \square$$

4.50 Remark. Since the notion of boundary is defined in terms of closure we see from Remark 4.30 that the boundary of a set is invariant under equivalent metrics. $\square$

4.51 Exercises.

1.   (i)   Consider the subset A of the closed interval $[0,1]$ consisting
               of those real numbers which have a decimal representation of
               the form

$$\sum_{n=1}^{\infty} \frac{a_n}{10^n} \quad \text{where } a_n \in \{0,1\} \text{ for all } n \in \mathbb{N}.$$

In $\mathbb{R}$ with the usual norm, determine whether

(a) $\cdot 11$ (b) $\cdot \ddot{0}\ddot{1}$ (c)   1

are cluster points of A.

(ii)   Consider the subset A of $\mathbb{R}^2$,

$$\left\{ (t, \sin \tfrac{1}{t}) : t \in (0, \tfrac{1}{\pi}) \right\}.$$

In $(\mathbb{R}^2, \| \cdot \|_2)$, determine whether

(a) $(\tfrac{1}{\pi}, 0)$ (b) $(0, 0)$ (c) $(1, 0)$

are cluster points of A.

(iii)   Consider the subset A of E_0 consisting of those sequences
         whose last non-zero term is strictly positive.
         In $(E_0, \| \cdot \|_\infty)$, determine whether

(a) $\{0, 0, \ldots\}$ (b) $\{0, 1, 0, \ldots\}$ (c) $\{0, -1, 0, \ldots\}$

are cluster points of A.

(iv)   Consider the subset A of $C[0,1]$ consisting of positive functions
         f where $f(0) = f(1) = 0$.
         In $(C[0,1], \| \cdot \|_1)$, determine whether

(a) $f(t) = 0$ (b) $f(t) = 1$ (c) $f(t) = \sin 2\pi t$

are cluster points of A.

2.   In the spaces given, determine the cluster points of the
following subsets.

    (i)   In $\mathbb{R}$ with the usual norm,

        (a) $\{(\frac{2^n-1}{2^n}, \frac{2^{n+1}-1}{2^{n+1}}) : n \in \mathbb{N}\}$

        (b) the set of dyadic fractions; that is,

            $\{\frac{a}{2^n} : a \in \mathbb{Z}, n \in \mathbb{N}\}$

    (ii)   In $\mathbb{C}$ with the usual norm,

        (a) $\{z \in \mathbb{C} : \text{Re } z \in \mathbb{Q}\}$

        (b) the set of all nth roots of unity for all $n \in \mathbb{N}$; that is,

            $\{z \in \mathbb{C} : e^{\frac{2k\pi i}{n}}, k \in \{1,2,\ldots,n\} \text{ and } n \in \mathbb{N}\}$.

    (iii)   In $(\mathbb{R}^2, \|\cdot\|_2)$,

        (a) $\{((-1)^n\frac{1}{n}, (-1)^n) : n \in \mathbb{N}\}$

        (b) $\{(\lambda,\mu) \in \mathbb{R}^2 : \mu = \lambda \quad \text{rational } \lambda$
            $= 1-\lambda \text{ irrational } \lambda\}$

3. (i)   Consider the set $\{f_n : n \in \mathbb{N}\}$ of real functions on $[0,1]$ where
            f_n is defined by

            $f_n(t) = t^n$.

            Determine the cluster points of this set in

        (a) $(\mathcal{B}[0,1], \|\cdot\|_\infty)$ and

        (b) $(C[0,1], \|\cdot\|_1)$.

    (ii)   Consider the set of graphs $\{G_{f_n} : n \in \mathbb{N}\}$ of the functions f_n
            in (i) where

            $G_{f_n} \equiv \{(t, f_n(t)) : t \in [0,1]\}$.

            Determine the cluster points of the union

            $\cup\{G_{f_n} : n \in \mathbb{N}\}$ in $(\mathbb{R}^2, \|\cdot\|_2)$.

4.   Using the uniform continuity of continuous functions on a bounded closed interval, prove that every continuous function on $[0,1]$ is a cluster point of the set of step functions on $[0,1]$ in the normed linear space $(\mathcal{B}[0,1]; \|\cdot\|_\infty)$; (This is the fundamental property used to establish the integrability of continuous functions.).

5.   In the spaces given, determine whether the following subsets are closed.

    (i)   In $\mathbb{R}$ with the usual norm,

       (a) $\mathbb{N}$,

       (b) the set of rational numbers which have a terminating decimal representation,

       (c) the set of real numbers which have a decimal representation consisting only of digits 0 and 9.

    (ii)   In $(\mathbb{R}^3, \|\cdot\|_2)$,

       (a) $\{x \in \mathbb{R}^3 : \|x\|_\infty \leqslant 1\}$,

       (b) $\{(\lambda, \mu, \nu) \in \mathbb{R}^3 : \lambda + \mu + \nu = 0 \text{ and } |\lambda|, |\mu|, |\nu| \leqslant 1\}$

       (c) $\{x \in \mathbb{R}^3 : \|x\| = 1\}$ for any norm $\|\cdot\|$ on $\mathbb{R}^3$.

    (iii)   In $(m, \|\cdot\|_\infty)$,

       (a) the set of sequences with second term zero,

       (b) the set of sequences whose terms consist only of digits 0 and 1,

       (c) the set of sequences each of whose terms is greater than or equal to 0 but less than or equal to 1.

    (iv)   In $(\mathcal{C}[-1,1], \|\cdot\|_\infty)$,

       (a) the set of functions f with the property that for a given $t_0 \in [-1,1]$, $|f(t_0)| \leqslant 1$,

       (b) the set of functions f with the property that there exists a $K > 0$ such that for each $f \in A$,

$$|f(s) - f(t)| \leqslant K|s-t| \quad \text{for all } s, t \in [-1,1].$$

       (c) the set of differentiable functions f with $f'(0) = 0$.

6.   Consider the real numbers in the interval $[0,1]$ represented in
ternary form; that is, for each $x \in [0,1]$

$$x = \sum_{k=1}^{\infty} \frac{a_k}{3^k} \quad \text{where } a_k \in \{0,1,2\} \text{ for each } k \in \mathbb{N}.$$

Cantor's Ternary set T consists of those real numbers in $[0,1]$ which have
a ternary representation of the form

$$x = \sum_{k=1}^{\infty} \frac{a_k}{3^k} \quad \text{where } a_k \in \{0,2\} \quad \text{for each } k \in \mathbb{N}.$$

(Note that $\frac{1}{3} = \cdot 1 \equiv \cdot 0\dot{2}$ (ternary), so $\frac{1}{3} \in T$.)
The defining property for a number to be an element of T is that it have
one ternary representation by 0s and 2s.)

Prove that   (i)   T is closed and

             (ii) every element of T is a cluster point of T.

(A set which has these two properties is called a *perfect* set.)

7.   (i)   Consider X a non-empty set with metrics d and d' with the
             property that any sequence $\{x_n\}$ convergent to x in (X,d) is
             convergent to x in (X,d').
             (a) Prove that for any subset A of X, the cluster points of A
                  in (X,d) are cluster points of A in (X,d').
             (b) Prove that if a subset A is closed in (X,d') then A is
                  closed in (X,d).
             (c) Deduce that a closed subset of a metric space is closed
                  with respect to any equivalent metric.

      (ii)   (a) Prove that if A is a closed subset of $(C[0,1], \|\cdot\|_1)$ then
                  A is a closed subset of $(C[0,1], \|\cdot\|_\infty)$.
             (b) Given an example to show that the converse does not hold
                  in general.

      (iii) Denote by $C^\infty[-1,1]$ the linear space of infinitely often
             differentiable functions on $[-1,1]$.
             Show that $C^\infty[-1,1]$ is not a closed linear subspace of
             $(C[-1,1], \|\cdot\|_\infty)$ and deduce that it is not a closed linear
             subspace of $(C[-1,1], \|\cdot\|_1)$.

8.   (i)   Given a metric space (X,d) and a subset Y of X, prove that a subset A of Y is closed in $(Y,d|_Y)$ if and only if there exists a closed set F in X such that $A = F \cap Y$.

(ii)   In the spaces given determine whether the following subsets A are closed in the specified subspaces Y.

(a)   In $\mathbb{R}$ with the usual norm,

$A \equiv \{\frac{1}{n} : n \in \mathbb{N}\}$ in $Y \equiv (0,1)$, and

$A \equiv \{$the set of real numbers which have terminating decimal representation$\}$ in $Y \equiv \mathbb{Q}$.

(b)   In $(E_0, \|\cdot\|_\infty)$,

$A \equiv \{$the set of sequences which have only one non-zero and strictly positive term$\}$ in

$Y_1 \equiv \{$the set of sequences which have only positive terms$\}$, and also in

$Y_2 \equiv \{$the set of sequences which have last non-zero term strictly positive$\}$.

(c)   In $(C[0,1], \|\cdot\|_\infty)$, given $f_0 \in C[0,1]$

$A \equiv \{$the set of polynomials p where $\|p-f_0\|_\infty \leqslant 1\}$ in $Y \equiv P$ the linear subspace of polynomials, and

$A \equiv \{$the set of cubic polynomials p where $\|p-f_0\|_\infty < 1\}$ in $Y \equiv B(0;1)$ the open unit ball.

9.   (i)   (a)   Denote by $B_0[0,1]$ the set of bounded real functions f on $[0,1]$ with the property that $f(0) = 0$. Prove that $B_0[0,1]$ is a closed linear subspace of $(B[0,1], \|\cdot\|_\infty)$ and deduce that $(B_0[0,1], \|\cdot\|_\infty)$ is a Banach space.

(b)   Denote by $C_0[0,1]$ the set of continuous functions f on $[0,1]$ with the property that $f(0) = 0$. Deduce from (i), or otherwise prove, that $C_0[0,1]$ is a closed linear subspace of $(C[0,1], \|\cdot\|_\infty)$.

(ii)   Prove that the linear subspace $C_0[0,1]$ is dense in $(C[0,1], \|\cdot\|_1)$.

10. (i) For the Banach space $(m, \|\cdot\|_\infty)$, prove that

    (a) c_0 is a closed linear subspace, and
    (b) c is a closed linear subspace.

    (ii) Deduce that

    (a) both $(c_0, \|\cdot\|_\infty)$ and $(c, \|\cdot\|_\infty)$ are Banach spaces and
    (b) c_0 is a closed linear subspace of $(c, \|\cdot\|_\infty)$.

    (iii) Show that the normed linear space $(\ell_1, \|\cdot\|_\infty)$ is not complete but prove that it is dense in $(c_0, \|\cdot\|_\infty)$.

    (iv) Show that the normed linear spaces $(E_0, \|\cdot\|_2)$ and $(\ell_1, \|\cdot\|_2)$ are not complete but prove that both are dense in $(\ell_2, \|\cdot\|_2)$.

11. Consider a metric space (X,d).

    (i) For any subset A prove that the derived set A' is closed.

    (ii) For subsets A and B, prove that

    (a) if $A \subseteq B$ then $\overline{A} \subseteq \overline{B}$, and
    (b) $\overline{A \cup B} = \overline{A} \cup \overline{B}$.

    (iii) Deduce from (i) that $\overline{A}$ is closed.

12. Consider a subset A of a metric space (X,d).

    (i) Prove that $x \in C(A)$ is a cluster point of A if and only if $d(x,A) = 0$.

    (ii) Prove that $x \in \partial A$ if and only if $d(x,A) = 0$ and $d\big(x,C(A)\big) = 0$.

13. Given a metric space (X,d) with subsets A and B where $A \subseteq B$.

    (i) Prove that if A is dense in $(B,d|_B)$ and B is dense in (X,d) then A is dense in (X,d).

    (ii) Prove that the normed linear space $(E_0, \|\cdot\|_2)$ is separable.

    (iii) Hence, or otherwise, prove that Hilbert sequence space $(\ell_2, \|\cdot\|_2)$ is separable.

14. (i)   Prove that a discrete metric space (X,d) is separable if and
            only if X is countable.

     (ii)   Given a non-empty set X, prove that the Banach space
            $(B(X), \|\cdot\|_\infty)$ is separable if and only if X is finite.

15. (i)   For any proper closed linear subspace M of a normed linear
            space $(X, \|\cdot\|)$, prove that

            (a) $\partial M = M.$ (b) $C(M)$ is dense in X.

     (ii)   Recalling that c denotes the linear space of all convergent
            sequences, prove that the subset of those which do not
            converge to 0 is dense in $(c, \|\cdot\|_\infty)$.

16. (i)   In a normed linear space $(X, \|\cdot\|)$, prove that the closure $\overline{A}$ of
            a convex set A is also convex.

     (ii)   In $(m, \|\cdot\|_\infty)$, prove that

            $$\overline{B[\underline{0};1] \cap E_0} = B[\underline{0};1] \cap c_0.$$

17.   *A normed algebra* $(A, \|\cdot\|)$ is an algebra A which is a normed
linear space where the norm $\|\cdot\|$ has the property

            $\|xy\| \leq \|x\| \|y\|$ for all $x,y \in A$
            (the submultiplicative property).

     (i)   Show that the normed linear space $(B(X), \|\cdot\|_\infty)$ is actually a
            normed algebra under pointwise definition of multiplication;
            that is,

            $fg(x) = f(x)g(x)$ for all $x \in X$.

     (ii)   Prove that in a normed algebra $(A, \|\cdot\|)$, the closure $\overline{B}$ of a
            subalgebra B is also an algebra.

     (iii) Consider the normed algebra $(B[a,b], \|\cdot\|_\infty)$.

            (a) Prove that the linear subspace $C[a,b]$ of continuous
                 functions on $[a,b]$ is a closed subalgebra.

17.(iii)   (continued)

        (b) Show that the linear subspace $P[a,b]$ of polynomials on [a,b] is not a closed subalgebra.

18.   A point x_0 of a subset E of a linear space X is said to be an *extreme point* of E if it cannot be expressed in the form $x_0 = \lambda x + (1-\lambda)y$ for any $x,y \in E$, $x \neq y$ and $0 < \lambda < 1$.

      (i)   Prove that the extreme points of a subset E of a normed linear space $(X, \|\cdot\|)$ are boundary points of E.
But show that in general not all boundary points of E are extreme points of E.

     (ii)   Determine the set of extreme points for the closed unit ball in the real Banach spaces

          (a) $(\ell_1, \|\cdot\|_1)$ and   (b) $(C[0,1], \|\cdot\|_\infty)$.

    (iii)   Show that the set of extreme points of any closed convex subset in $(\mathbb{R}^n, \|\cdot\|_2)$ when $n \geq 3$, is not necessarily closed. Is this also the case when $n = 2$?

19.   Consider a linear space X with linear subspaces M and N such that $X = M \oplus N$.

      (i)   Prove that if $\|\cdot\|$ is a norm for X then $\|\cdot\|'$ defined by

          $\|x\|' = \|m\| + \|n\|$ where $x = m+n$ and $m \in M$, $n \in N$,

          is also a norm for X.

     (ii)   Prove that if $(X, \|\cdot\|)$ is complete and M and N are closed subspaces of $(X, \|\cdot\|)$ then $(X, \|\cdot\|')$ is also complete.

20.   A sequence $\{E_n\}$ of subsets of a metric space (X,d) is said to
be *decreasing* if $E_1 \supseteq E_2 \supseteq \ldots \supseteq E_n \supseteq \ldots$ Prove that a metric space (X,d)
is complete if and only if for every decreasing sequence $\{F_n\}$ of non-empty
closed sets such that $\delta(F_n) \to 0$, we have $\cap\{F_n : n \in \mathbb{N}\}$ contains exactly
one point.

(This is called *Cantor's Intersection Theorem* and it provides an alternative
characterisation of completeness.

Hint: Given (X,d) complete and a decreasing sequence $\{F_n\}$, choose a
sequence $\{x_n\}$ such that $x_n \in F_n$ for each $n \in \mathbb{N}$.

            Conversely, given a Cauchy sequence $\{x_n\}$, consider the
sequence $\{F_n\}$ of sets where $F_n \equiv \{x_k : k \geq n\}$ for each $n \in \mathbb{N}$.)

21. (i)   For a normed linear $(X, \|\cdot\|)$ with proper closed linear subspace
            M, prove that a norm $\|\cdot\|'$ is defined on the quotient space $\frac{X}{M}$ by

            $\|[x]\|' = \inf\{\|x+m\| : m \in M\}$ for any $x \in [x]$.

     (ii)   Prove that if $(X, \|\cdot\|)$ is a Banach space then $(\frac{X}{M}, \|\cdot\|')$ is also
            a Banach space.

5. APPLICATION: BANACH'S FIXED POINT THEOREM

One of the most impressive applications of the theory we have
so far developed is in Banach's Fixed Point Theorem sometimes called
Banach's Contraction Mapping Principle.

This theorem shows the underlying importance of completeness
in providing an existence result. But remarkably, this theorem also
provides a method for deriving the object whose existence is guaranteed.
So the theorem is at the same time both theoretical and practical, a rare
phenomenon indeed!

The theorem is set in the context of complete metric spaces,
so it is widely applicable to a number of situations which at first sight
seem unrelated. This demonstrates the power of the method of developing
a general theory for abstract metric spaces as a unifying technique to
comprehend a variety of problem areas.

We will see that the theorem is used in real analysis, numerical
analysis and the theory of ordinary differential and integral equations.

5.1 Definition. Given a non-empty set X, an element $x \in X$ is called
a fixed point of a mapping T of X into itself if $Tx = x$.

5.2 Remark. A mapping T from X into itself may have no, one or many fixed
points:
Consider $\mathbb{R}^2$.

    (i)   The translation mapping $T : \mathbb{R}^2 \to \mathbb{R}^2$ where given a $\neq 0$,
               $Tx = a+x$, has no fixed point.

    (ii)   The rotation mapping $R_0 : \mathbb{R}^2 \to \mathbb{R}^2$ where given $0 < \alpha < 2\pi$,

$$R_0(\lambda,\mu) = \begin{bmatrix} \cos\alpha & -\sin\alpha \\ \sin\alpha & \cos\alpha \end{bmatrix} \begin{pmatrix} \lambda \\ \mu \end{pmatrix},$$

           has only one fixed point, $(0,0)$.

    (iii)   The reflection mapping $R_e : \mathbb{R}^2 \to \mathbb{R}^2$ where $R_e(\lambda,\mu) = (\lambda,-\mu)$,
           has every point of $\{(\lambda,0) : \lambda \in \mathbb{R}\}$ as a fixed point. $\square$

The fixed points of a mapping T from X into itself are in fact
solutions of the equation

$$(T-I)(x) = 0$$

where I is the identity mapping on X. So a standard technique for finding
solutions of an equation is to find the fixed points of an associated
mapping.

5.3 Definition. Given a metric space (X,d) a mapping T of X into itself is
called a *contraction mapping* if there exists an $0 < \alpha < 1$ such that

$$d(Tx,Ty) \leq \alpha d(x,y) \quad \text{for all } x,y \in X.$$

Banach's Fixed Point Theorem assures us of the existence and
uniqueness of a fixed point for any contraction mapping on a complete
metric space and provides a method to compute it.

5.4 Banach's Fixed Point Theorem.

(i)   *A contraction mapping T of a complete metric space (X,d) into*
      itself has one and only one fixed point.

(ii) *Given any $x_0 \in X$ the sequence of iterations $\{T^n x_0\}$ is*
      convergent to that fixed point of T.

Proof. Given $x_0 \in X$, consider the sequence $x_1 \equiv Tx_0$,
$x_2 \equiv Tx_1 = T^2 x_0, \ldots, x_n \equiv T^n x_0, \ldots$ of iterations.
For $m > n$,

$$
\begin{aligned}
d(x_n, x_m) &= d(T^n x_0, T^m x_0) = d(T^n x_0, T^n T^{m-n} x_0) \\
&\leq \alpha^n d(x_0, T^{m-n} x_0) = \alpha^n d(x_0, x_{m-n}) \\
&\leq \alpha^n \{ d(x_0, x_1) + d(x_1, x_2) + \ldots + d(x_{m-n-1}, x_{m-n}) \} \\
&\leq \alpha^n d(x_0, x_1) \{ 1 + \alpha + \alpha^2 + \ldots + \alpha^{m-n-1} \} \\
&< \frac{\alpha^n}{1-\alpha} d(x_0, x_1). \qquad\qquad \ldots \text{(*)}
\end{aligned}
$$

Since $0 < \alpha < 1$, we conclude that $\{x_n\}$ is a Cauchy sequence. But since
(X,d) is complete there exists an $x \in X$ to which $\{x_n\}$ is convergent.
We show that x is a fixed point of T:

Now $\qquad d(Tx,x) \leq d(Tx, T^n x_0) + d(T^n x_0, x)$

$$\qquad\qquad \leq \alpha d(x, x_{n-1}) + d(x_n, x).$$

and since $\{x_n\}$ is convergent to x we deduce that $d(Tx,x) = 0$ which implies that $Tx = x$.

We show that x is a unique fixed point of T:

Suppose that for some $y \in X$, $Ty = y$.

Then $d(x,y) = d(Tx,Ty) \leqslant \alpha d(x,y)$.

Since $0 < \alpha < 1$ then $d(x,y) = 0$ which implies that $y = x$. $\square$

5.5 Remark. It follows from (∗) that

$$d(x_n,x) \leqslant \frac{\alpha^n}{1-\alpha} d(x_0,x_1) \qquad \qquad \ldots \text{ (i)}$$

and this can be used to estimate the error in choosing the nth iterate as an approximation to the fixed point. However, this can be very rough because the choice of starting point for the iteration may produce an iteration sequence which is initially slow to converge.

A better estimate of error is given by

$$d(x_n,x) \leqslant \frac{\alpha}{1-\alpha} d(x_{n-1},x_n) \qquad \qquad \ldots \text{ (ii)}$$

which is found from (i) by considering the case n = 1, which gives

$$d(x_1,x) \leqslant \frac{\alpha}{1-\alpha} d(x_0,x_1)$$

and by considering the iteration as starting from x_{n-1}; that is, replacing x_0 by x_{n-1} and x_1 by x_n. $\square$

5.6 Remark. In particular situations where we would like to apply Banach's Fixed Point Theorem it often happens that the mapping T is not a contraction over the whole space (X,d) but only on a subset Y of it. If the subset Y is closed then by Theorem 4.20(ii), $(Y,d|_Y)$ is complete so if T maps Y into Y then we can apply Banach's Fixed Point Theorem. $\square$

5.7 Application in real analysis.

A real function f mapping the closed interval [a,b] into itself and satisfying the Lipschitz condition

Limit processes

$$|f(x)-f(y)| \leq K|x-y| \qquad \text{for all } x,y \in [a,b]$$

where $0 < K < 1$, is a contraction mapping. So for any $x_0 \in [a,b]$ the sequence of iterations $x_0, x_1 \equiv f(x_0), \ldots, x_n \equiv f(x_{n-1}), \ldots$ converges to the unique fixed point of f.

In particular, if f is differentiable on $[a,b]$ and

$$|f'(x)| \leq K < 1 \qquad \text{for all } x \in [a,b]$$

then the Mean Value Theorem for differential calculus implies that f satisfies such a Lipschitz condition.

It is instructive to visualise the successive iterations for two different cases:

(i)

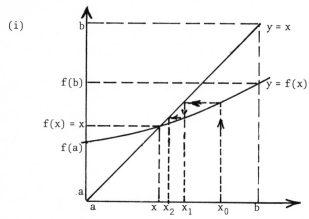

Figure 9.   Iteration when $0 < f'(x) \leq K < 1$ for all $x \in [a,b]$

(ii)

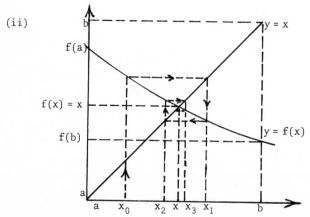

Figure 10.   Iteration when $-1 < -K \leq f'(x) < 0$ for all $x \in [a,b]$.

Consider the problem of solving the equation $f(x) = 0$ where $f(a) < 0$ and $f(b) > 0$ and f is differentiable on $[a,b]$ and $0 < k \leqslant f'(x) \leqslant K$ for all $x \in [a,b]$.

We may choose $\lambda > 0$, small enough so that the associated function g on $[a,b]$ defined by

$$g(x) = x - \lambda f(x)$$

maps into $[a,b]$. But also

$$1 - \lambda K \leqslant g'(x) \leqslant 1 - \lambda k \quad \text{for all } x \in [a,b]$$

so choosing $\lambda < \frac{2}{K}$ we see that g is a contraction mapping and the fixed point of g is the solution of the equation $f(x) = 0$.

5.7.1 Example. Solving the equation

$$x^3 + x - 1 = 0$$

using this method we begin by noting that the equation has one real root between 0 and 1, and writing

$$f(x) = x^3 + x - 1$$

we see that

$$1 \leqslant f'(x) \leqslant 4 \quad \text{for all } x \in [0,1].$$

The associated function g on $[0,1]$ is defined by

$$g(x) = x - \lambda(x^3 + x - 1) \quad \text{for } 0 < \lambda \leqslant 1.$$

Now $\qquad 1 - 4\lambda \leqslant g'(x) \leqslant 1 - \lambda \quad \text{for all } x \in [0,1]$

so choosing $\lambda = \frac{1}{4}$ we have $|g'(x)| \leqslant \frac{3}{4}$.

Now $\qquad g(x) = -\frac{1}{4}(x^3 - 3x - 1)$.

For $x_0 = \frac{1}{2}$, $x_1 \equiv g(\frac{1}{2}) = \frac{19}{32}$, $x_2 = 0 \cdot 643\ldots$. The error after n iterations is by (i) Remark 5.5, less than or equal to $(\frac{3}{4})^n \cdot \frac{3}{8}$ which shows that the convergence of the iterations could be made more rapid by a better choice of λ closer to $\frac{1}{2}$.

An alternative method would follow from noting that the
function h on $[0,1]$ defined by

$$h(x) = \frac{1}{1+x^2}$$

is a contraction mapping on $[0,1]$ and the fixed point of h is the solution
to the equation. □

5.7.2 Remark. Some would be familiar with Newton's method for finding a
simple root of the equation $f(x) = 0$ where f is differentiable on an
interval containing the root. It is of interest to see that for the
case where f has continuous second derivative f" on the interval then
Newton's method is justified by being an application of Banach's Fixed
Point Theorem.

In Newton's method the associated function g is defined by

$$g(x) = x - \frac{f(x)}{f'(x)} \ .$$

Suppose there is a simple root in the interval $[a,b]$ and $f(a) < 0$ and
$f(b) > 0$ and $f'(x) > 0$ for all $x \in [a,b]$. Then interval $g([a,b])$ contains
the root.

Now $$g'(x) = \frac{f(x)}{(f'(x))^2} \ f''(x) \ .$$

Since f' and f" are bounded on $[a,b]$ and f is continuous and is zero at
the required root, we can adjust the interval $[a,b]$ around the root so
that g is a contraction mapping on the adjusted interval $[a,b]$. □

5.8 Application in linear algebra.

It is interesting to see that Banach's Fixed Point Theorem can
be applied to determine approximations to the solution of a system of
linear equations , a situation which is an algebraic problem and which
does not involve a metric at all!

The system of n linear equations

$$\mu_i = \sum_{j=1}^{n} a_{ij}\lambda_j + \beta_i \qquad i \in \{1,2,\dots,n\}$$

can be thought of as a mapping $T: \mathbb{R}^n \to \mathbb{R}^n$

$$y = Tx = Ax + b$$

where $x \equiv (\lambda_1, \lambda_2, \ldots, \lambda_n)$

              $y \equiv (\mu_1, \mu_2, \ldots, \mu_n)$

and $b \equiv (\beta_1, \beta_2, \ldots, \beta_n)$

and   A is the $n \times n$ matrix (a_{ij}).

Let us consider the problem of finding a fixed point for the mapping T.

The Banach Fixed Point Theorem requires a norm for $\mathbb{R}^n$. From Theorem 3.36 we have that $\mathbb{R}^n$ with any norm is complete. Whether or not T is a contraction mapping on $\mathbb{R}^n$ depends on our choice of norm for $\mathbb{R}^n$. We now determine sufficient conditions for T to be a contraction mapping under different choices of norm for $\mathbb{R}^n$.

Case 1.   Under the Euclidean norm $\|\cdot\|_2$.

Now $\|y-y'\|_2^2 = \|A(x-x')\|_2^2$

$$= \sum_{i=1}^{n} \left(\sum_{j=1}^{n} a_{ij}(\lambda_j - \lambda_j') \right)^2$$

But $\sum_{j=1}^{n} a_{ij}(\lambda_j - \lambda_j') \le \left(\sum_{j=1}^{n} a_{ij}^2 \right)^{\frac{1}{2}} \left(\sum_{j=1}^{n} (\lambda_j - \lambda_j')^2 \right)^{\frac{1}{2}}$

by the Cauchy-Schwarz inequality

so $\|y-y'\|_2^2 \le \left(\sum_{i=1}^{n} \sum_{j=1}^{n} a_{ij}^2 \right) \|x-x'\|_2^2$.

We conclude that if $\sum_{i=1}^{n} \sum_{j=1}^{n} a_{ij}^2 < 1$ then T is a contraction mapping on $(\mathbb{R}^n, \|\cdot\|_2)$. $\square$

The conditions for T to be a contraction mapping may be less constrained under a different choice of norm.

Case 2. Under the $\|\cdot\|_\infty$-norm.

Now
$$\|y-y'\|_\infty = \|A(x-x')\|_\infty$$

$$\leq \max_i \left(\sum_{j=1}^n |a_{ij}| |\lambda_j - \lambda_j'| \right)$$

$$\leq \left(\max_i \sum_{j=1}^n |a_{ij}| \right) \left(\max_j |\lambda_j - \lambda_j'| \right)$$

$$= \left(\max_i \sum_{j=1}^n |a_{ij}| \right) \|x-x'\|_\infty .$$

We conclude that if $\sum_{j=1}^n |a_{ij}| < 1$ for each $i \in \{1,2,\ldots,n\}$ then T is a contraction mapping on $(\mathbb{R}^n, \|\cdot\|_\infty)$ which is certainly a less constrained condition than that from Case 1. $\square$

Case 3. Under the $\|\cdot\|_1$-norm.

Now
$$\|y-y'\|_1 = \|A(x-x')\|_1$$

$$= \sum_{i=1}^n \left| \sum_{j=1}^n a_{ij} (\lambda_j - \lambda_j') \right|$$

$$\leq \sum_{i=1}^n \sum_{j=1}^n |a_{ij}| |\lambda_j - \lambda_j'|$$

$$\leq \left(\max_j \sum_{i=1}^n |a_{ij}| \right) \|x-x'\|_1 .$$

We conclude that if $\sum_{i=1}^n |a_{ij}| < 1$ for each $j \in \{1,2,\ldots,n\}$ then T is a contraction mapping on $(\mathbb{R}^n, \|\cdot\|_1)$ which is a condition complementary to that from Case 2 and also less constrained than that from Case 1. $\square$

Each of the Cases 1, 2 and 3 provides sufficient but not necessary conditions to apply the iterative procedure of Banach's Fixed Point Theorem , an algebraic procedure , to find the fixed point.

We remind ourselves that the fixed point of the system

$$y = Tx = Ax + b$$

is the solution of the system

$$(A-I)x + b = 0.$$

It might be asked why such a procedure should be used when there are more direct methods of solving such equations. We have from the three cases, easily applicable tests which indicate the invertibility of the matrix $A - I$ and with large sparse matrices the iterative process can have computational advantages in producing approximate solutions.

In numerical analysis the Jacobi and the Gauss-Seidel iteration procedures are practical modifications of this basic idea.

<u>5.8.1 Example.</u> We solve the system of equations

$$4\lambda_1 - \lambda_2 - \lambda_3 \qquad = 2$$

$$-\lambda_1 + 4\lambda_2 \qquad - \lambda_4 = 2$$

$$-\lambda_1 \qquad + 4\lambda_3 - \lambda_4 = 1$$

$$\qquad - \lambda_2 - \lambda_3 + 4\lambda_4 = 1$$

using the iteration method:

Writing

$$A = \begin{pmatrix} 4 & -1 & -1 & 0 \\ -1 & 4 & 0 & -1 \\ -1 & 0 & 4 & -1 \\ 0 & -1 & -1 & 4 \end{pmatrix} \quad \text{and} \quad b = \begin{pmatrix} 2 \\ 2 \\ 1 \\ 1 \end{pmatrix},$$

the solution is the fixed point of the mapping T on $\mathbb{R}^4$ defined by

$$Tx = (I - \tfrac{1}{4}A)x + \tfrac{1}{4}b.$$

The rescaling of the original system provides a simple matrix $I - \tfrac{1}{4}A$ which satisfies either of the conditions given in Cases 2 and 3 to provide that T is a contraction mapping.

So

$$T\begin{pmatrix} \lambda_1 \\ \lambda_2 \\ \lambda_3 \\ \lambda_4 \end{pmatrix} = \tfrac{1}{4}\begin{pmatrix} 0 & 1 & 1 & 0 \\ 1 & 0 & 0 & 1 \\ 1 & 0 & 0 & 1 \\ 0 & 1 & 1 & 0 \end{pmatrix}\begin{pmatrix} \lambda_1 \\ \lambda_2 \\ \lambda_3 \\ \lambda_4 \end{pmatrix} + \begin{pmatrix} \tfrac{1}{2} \\ \tfrac{1}{2} \\ \tfrac{1}{4} \\ \tfrac{1}{4} \end{pmatrix}$$

Beginning the iteration from $x_0 \equiv (1,1,1,1)$ we have $x_1 \equiv (1,1,\tfrac{3}{4},\tfrac{3}{4})$, $x_2 \equiv (\tfrac{15}{16},\tfrac{15}{16},\tfrac{11}{16},\tfrac{11}{16})$, $x_3 \equiv (\tfrac{29}{32},\tfrac{29}{32},\tfrac{21}{32},\tfrac{21}{32})$, which compares with the exact solution $x \equiv (\tfrac{7}{8},\tfrac{7}{8},\tfrac{5}{8},\tfrac{5}{8})$. □

5.9 Application in the theory of differential equations.

To prove the existence and uniqueness of solutions of some types of differential equations, Banach's Fixed Point Theorem is applied to subspaces of the function space $(C[a,b], \|\cdot\|_\infty)$.

5.9.1 Picard's Theorem. *Consider the differential equation*

$$\frac{dy}{dx} = f(x,y)$$

with the initial condition

$$y(x_0) = y_0$$

where f is defined on a closed rectangle $D \equiv \{(x,y) : |x-x_0| \le a, \; |y-y_0| \le b\}$ *in* $\mathbb{R}^2$, *and is continuous on* D *and satisfies a Lipschitz condition with respect to y on* D; *that is, there exists an* $M > 0$ *such that*

$$|f(x,y_1)-f(x,y_2)| \le M|y_1-y_2| \quad \text{for all } (x,y_1),(x,y_2) \in D.$$

Then on some closed interval $[x_0-r, x_0+r]$ *where* $0 < r < a$, *there exists a unique solution satisfying the initial condition.*

Proof. The differential equation with the initial condition is equivalent to the integral equation

$$y(x) = y_0 + \int_{x_0}^{x} f(t, y(t)) \, dt.$$

Since f is continuous on the bounded closed rectangle D, there exists a $k > 0$ such that

$$|f(x,y)| \le k \quad \text{for all } (x,y) \in D.$$

Choose $0 < r < \min\{\frac{1}{M}, a, \frac{b}{k}\}$. Consider the Banach space $(C[x_0-r, x_0+r], \|\cdot\|_\infty)$ and the constant function y_0 where

$$y_0(x) = y_0 \quad \text{for all } x \in [x_0-r, x_0+r].$$

Now the set of continuous functions $y(x)$ on $[x_0-r,x_0+r]$ such that

$$|y(x)-y_0| \leq kr \quad \text{for all } x \in [x_0-r,x_0+r],$$

is the closed ball $B[y_0(x);kr]$ in $(C[x_0-r,x_0+r],\|\cdot\|_\infty)$ so $B[y_0(x);kr]$ with metric induced by $\|\cdot\|_\infty$ is a complete metric space.

Consider the mapping T from $B[y_0(x);kr]$ into $C[x_0-r,x_0+r]$ defined by

$$Ty(x) = y_0 + \int_{x_0}^{x} f(t,y(t))dt.$$

Now T maps $B[y_0(x);kr]$ into itself since

$$|Ty(x)-y_0| \leq \int_{x_0}^{x} |f(t,y(t))|dt$$

$$\leq k|x-x_0| \leq kr \quad \text{for all } x \in [x_0-r,x_0+r].$$

But also T is a contraction mapping since

$$|Ty_1(x)-Ty_2(x)| \leq \int_{x_0}^{x} |f(t,y_1(t)) - f(t,y_2(t))|dt$$

$$\leq Mr\|y_1-y_2\|_\infty \quad \text{and} \quad Mr < 1.$$

It follows from Banach's Fixed Point Theorem that T has a unique fixed point; that is, there exists a unique continuous function $y(x)$ on $[x_0-r,x_0+r]$ such that

$$y(x) = y_0 + \int_{x_0}^{x} f(t,y(t))dt$$

or equivalently, since f is continuous, such that

$$\frac{dy}{dx} = f(x,y(x))$$

and $y(x_0) = y_0.$ $\square$

The solution of the differential equation is the limit function in the *Picard iteration* defined by

$$y_{n+1}(x) = y_0 + \int_{x_0}^{x} f(t,y_n(t))dt \quad \text{for } n \in \mathbb{Z}^+.$$

However, the practical usefulness of this iteration method in finding the solution is generally limited by the difficulty of the integrations involved.

It should be noted that if f has bounded partial derivative with respect to y on D then the Mean Value Theorem for differential calculus implies that f satisfies a Lipschitz condition with respect to y on D.

5.9.2 Example. We approximate the solution to the differential equation

$$\frac{dy}{dx} = x - y^2$$

with initial condition $y(0) = \frac{1}{2}$ using the Picard iteration:

Now $f(x,y) = x - y^2$

satisfies the conditions of Picard's Theorem. The Picard iteration is given by

$$
\begin{aligned}
y_1(x) &= \tfrac{1}{2} + \int_0^x (t-\tfrac{1}{4})dt \\
       &= \tfrac{1}{2} - \tfrac{x}{4} + \tfrac{x^2}{2} \, . \\
y_2(x) &= \tfrac{1}{2} + \int_0^x (t - \tfrac{1}{16}(2-t+2t^2)^2)dt \\
       &= \tfrac{1}{2} - \tfrac{x}{4} + \tfrac{5}{8}x^2 - \tfrac{3}{16}x^3 + \tfrac{1}{16}x^4 - \tfrac{1}{20}x^5 \, .
\end{aligned}
$$

It is evident that the labour in pursuing further iterations would soon become intolerable!

However, even at this stage we can give an estimate of error: If we suppose f defined on

$$D \equiv \{(x,y) : |x| \leqslant 1, \ |y-\tfrac{1}{2}| \leqslant \tfrac{1}{2}\},$$

then $|f(x,y)| \leqslant 1$ on D and M = 2 so $r < \frac{1}{2}$. From formula (i) in Remark 5.5 we have $\alpha = Mr < 1$,

$$\|y_2(x)-y(x)\|_\infty \leqslant \frac{\alpha^2}{1-\alpha} \|y_1(x)-y_0(x)\|_\infty$$

So, say on $[-\frac{3}{8},\frac{3}{8}]$, $\alpha = \frac{3}{4}$ and

$$\|y_2-y\|_\infty \leq \frac{9}{40} \cdot \frac{21}{128} < \frac{1}{25} \cdot \square$$

5.10 Application in the theory of integral equations.

The method of proof of Picard's Theorem, transforming a differential equation into an equivalent integral equation, suggests that Banach's Fixed Point Theorem can be used to prove the existence and uniqueness of solutions of some types of integral equations.

5.10.1 Fredholm integral equations of the second kind. Such are integral equations of the form

$$y(x) = v(x) + \lambda \int_a^b k(x,t)y(t)dt$$

where   v is a given continuous function on $[a,b]$

y is the required solution function on $[a,b]$

k, called the *kernel* of the equation, is a given continuous function on the square region $\square \equiv \{(x,t) : a \leq x \leq b, a \leq t \leq b\}$ and λ is a parameter.

Since k is continuous on the bounded closed square $\square$, there exists an $M > 0$ such that

$$|k(x,y)| \leq M \quad \text{for all } (x,y) \in \square .$$

Consider the mapping T from the Banach space $(C[a,b], \|\cdot\|_\infty)$ into itself defined by

$$Ty(x) = v(x) + \lambda \int_a^b k(x,t)y(t)dt.$$

Now $\|Ty_1-Ty_2\|_\infty \leq |\lambda|M(b-a)\|y_1-y_2\|_\infty$

so T is a contraction mapping if

$$|\lambda| < \frac{1}{M(b-a)} .$$

The solution of the Fredholm integral equation is the fixed point of the mapping T and we conclude from Banach's Fixed Point Theorem that, when $|\lambda| < \frac{1}{M(b-a)}$, there exists a unique fixed point in $C[a,b]$. The solution is the limit of the iteration

$$y_{n+1}(x) = v(x) + \lambda \int_a^b k(x,t)y_n(t)dt \quad \text{for } n \in \mathbb{Z}^+$$

beginning from an initial function y_0 on $[a,b]$.

5.10.2 Example.

We solve the Fredholm integral equation of the second kind

$$y(x) = 1 + \lambda \int_0^1 e^{x-t}y(t)dt$$

by iteration beginning with $y_0(x) = 1$:

Now $|e^{x-y}| \leqslant e$ for all $0 \leqslant x \leqslant 1$, $0 \leqslant y \leqslant 1$, so we solve for $|\lambda| < \frac{1}{e}$.

$$y_1(x) = 1 + \lambda \int e^{x-t}dt$$

$$= 1 - \lambda e(\tfrac{1}{e} - 1).$$

$$y_2(x) = 1 + \lambda \int_0^1 e^{x-t}\left(1 - \lambda e^t(\tfrac{1}{e} - 1)\right)dt$$

$$= 1 - \lambda e^x(\tfrac{1}{e} - 1)(1+\lambda)$$

$$\cdot \ \cdot \ \cdot$$

$$y_n(x) = 1 - \lambda e^x(\tfrac{1}{e} - 1)(1+\lambda+\ldots+\lambda^{n-2})$$

$$\to 1 - \frac{\lambda}{1-\lambda} e^x(\tfrac{1}{e} - 1) \quad \text{since } |\lambda| < \tfrac{1}{e} ,$$

and we check that this is indeed the solution. It is clear that this is a solution for $|\lambda| < 1$; the contraction mapping constraint on λ is a sufficient but not necessary condition for the iteration procedure to generate a solution. □

5.10.3 Volterra integral equations. Such are integral equations of the form

$$y(x) = v(x) + \lambda \int_a^x k(x,t)y(t)dt$$

where   v is a given continuous function on $[a,b]$

y is the required solution function on $[a,b]$

k, called the kernel of the equation, is a given continuous function on the triangular region $\Delta \equiv \{(x,t) : a \leqslant t \leqslant x, a \leqslant x \leqslant b\}$,

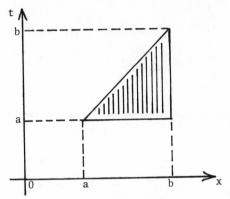

Figure 11.   The triangular region $\Delta \equiv \{(x,t) : a \leqslant t \leqslant x, a \leqslant x \leqslant b\}$.

and λ is a parameter.

Since k is continuous on the bounded closed triangle Δ, there exists an $M > 0$ such that

$$|k(x,t)| \leqslant M \quad \text{for all } (x,t) \in \Delta.$$

Consider the mapping T from the Banach space $(C[a,b], \|\cdot\|_\infty)$ into itself defined by

$$Ty(x) = v(x) + \lambda \int_a^x k(x,t)y(t)dt .$$

Now $\|Ty_1-Ty_2\|_\infty \leqslant |\lambda|M(x-a)\|y_1-y_2\|_\infty$

and by induction we establish that

$$\left\| T^m y_1 - T^m y_2 \right\|_\infty \leq |\lambda|^m M^m \frac{(x-a)^m}{m!} \left\| y_1 - y_2 \right\|_\infty \quad \text{for all } m \in \mathbb{N}.$$

Therefore,

$$\left\| T^m y_1 - T^m y_2 \right\|_\infty \leq |\lambda|^m M^m \frac{(b-a)^m}{m!} \left\| y_1 - y_2 \right\|_\infty \quad \text{for all } m \in \mathbb{N}.$$

For sufficiently large m,

$$|\lambda|^m M^m \frac{(b-a)^m}{m!} < 1$$

and so the corresponding T^m is a contraction mapping.

To proceed further we need the following general lemma which extends Banach's Fixed Point Theorem.

<u>5.10.4 Lemma</u>. *If* T *is a mapping from a complete metric space* (X,d) *into itself such that* T^m *is a contraction mapping for some* $m \in \mathbb{N}$ *then* T *has one and only one fixed point.*

<u>Proof</u>. Since T^m is a contraction mapping there exists an $0 < \alpha < 1$ such that

$$d(T^m x, T^m y) \leq \alpha d(x,y) \quad \text{for all } x,y \in X.$$

By Banach's Fixed Point Theorem, T^m has a unique fixed point, say x. Then since $T^m x = x$,

$$d(Tx,x) = d(T^{m+1}x, T^m x) \leq \alpha d(Tx,x)$$

so $d(Tx,x) = 0$ and $Tx = x$; that is, x is a fixed point of T. Since every fixed point of T is a fixed point of T^m it follows that T has one and only one fixed point. $\square$

The solution of the Volterra integral equation is the unique fixed point in $C[a,b]$ of the mapping T and we are assured from the lemma that this fixed point exists and is unique for all values of the parameter λ.

We notice that the Volterra integral equation can be regarded as a special type of Fredholm integral equation where the kernel k is zero on $\{(x,t) : x < t \leqslant b, \ a \leqslant x \leqslant b\}$ and may be discontinuous at points of the diagonal $\{(x,t) : t = x, \ a \leqslant x \leqslant b\}$.

5.11 Exercises.

1. (i)   Show that the cosine function on $\mathbb{R}$ is a contraction mapping on $[0,a]$ for any $1 \leqslant a < \frac{\pi}{2}$. Using Banach's Fixed Point Theorem find the solution to the equation

$$x = \cos x$$

correct to three decimal places.

(ii)   Show that the function f on $\mathbb{R}$ defined by

$$f(x) = \frac{\sin x}{x}, \ x \neq 0$$
$$f(0) = 1$$

is a contraction mapping on $[0,\pi]$.
Using Banach's Fixed Point Theorem find the non-zero solution to the equation

$$x^2 = \sin x$$

correct to three decimal places.

2. (i)   Show that the equation

$$xe^x = 1$$

has a root between $\frac{1}{e}$ and $1 - \frac{1}{e}$ and find this root correct to three decimal places.

(ii)   Find the value of x, correct to three decimal places, for which

$$\int_0^x \frac{t^2}{1+t^2} \, dt = \frac{1}{2} \ .$$

3.   Consider the statement of Banach's Fixed Point Theorem.

   (i)   Give an example of a contraction mapping of an incomplete
            metric space into itself which has no fixed point.

   (ii)   Give an example of a mapping T of a complete metric space into
            itself with the property

            $d(Tx,Ty) < d(x,y)$, for all x,y, $x \neq y$

            but which has no fixed point.

   (iii)   Give an example of a mapping T of a complete metric space into
            itself with the property that for some $m \in \mathbb{N}$, T^m is a
            contraction mapping but where T is not a contraction mapping.

4.   An iteration procedure for calculating the square root of a
given positive number α is

$$x_{n+1} = \frac{1}{2}(x_n + \frac{\alpha}{x_n}) \quad \text{for } n \in \mathbb{Z}^+$$

and where the sequence of iterates $\{x_n\}$ converges to $\sqrt{\alpha}$.

   (i)   Justify this iteration procedure by reference to Banach's
            Fixed Point Theorem.

   (ii)   Beginning from $x_0 = 1$, calculate an approximation for $\sqrt{2}$
            correct to three decimal places.

5.   (A special case of the *Implicit Function Theorem*)
Consider a real continuous function f on $D \equiv \{(x,y) \in \mathbb{R}^2 : a \leq x \leq b\}$
where the partial derivatives with respect to y exist and there exists
$m,M > 0$ such that

$$0 < m < \frac{\partial f}{\partial y} \leq M \quad \text{for all } (x,y) \in D.$$

Prove that there exists one and only one continuous function $y(x)$ on $[a,b]$
such that

$$f\big(x,y(x)\big) = 0.$$

[Hint: Consider the mapping T from $(C[a,b], \|\cdot\|_\infty)$ into itself defined by
$Ty(x) = y(x) - \frac{1}{M}f\big(x,y(x)\big)$.]

6.   Solve the system of equations

$$
\begin{bmatrix}
5 & -1 & -1 & -1 \\
-1 & 10 & -1 & -1 \\
-1 & -1 & 5 & -1 \\
-1 & -1 & -1 & 10
\end{bmatrix}
\underset{\sim}{x}
=
\begin{bmatrix}
-4 \\
12 \\
8 \\
34
\end{bmatrix}
$$

using the iteration method and beginning the iteration with $x_0 = (0,0,0,0)$. Show that the fifth iterate entries differ from the exact solution entries by less than $\cdot 06$.

7.   (Systems of *non-linear equations*)

(i)   Consider a system of m mappings $\{f_i : i \in \{1,2,\ldots,m\}\}$ for a bounded closed rectangular region D in $\mathbb{R}^m$ into $\mathbb{R}$, such that each f_i has continuous partial derivatives on D. Using the extended Mean Value Theorem for each f_i we have for any $x \equiv (\lambda_1,\lambda_2,\ldots,\lambda_n)$ and $x' \equiv (\lambda_1',\lambda_2',\ldots,\lambda_n')$ that

$$
f_i(x) - f_i(x') = (\lambda_1-\lambda_1')\frac{\partial f_i}{\partial \lambda_1}(\xi_i) + \ldots + (\lambda_m-\lambda_m')\frac{\partial f_i}{\partial \lambda_m}(\xi_i)
$$

where $\xi_i = \lambda_i x + (1-\lambda_i)x$ for some $0 < \lambda_i < 1$.

Writing $f_{ij} = \sup\limits_{x \in D}\left|\frac{\partial f_i}{\partial \lambda_j}\right|$ for $i,j \in \{1,2,\ldots,m\}$,

show that if the matrix $[f_{ij}]$ has any one of the properties

(a) $\sum\limits_{j=1}^{m} |f_{ij}| < 1$ for each $i \in \{1,2,\ldots,m\}$,

(b) $\sum\limits_{i=1}^{m} |f_{ij}| < 1$ for each $j \in \{1,2,\ldots,m\}$ or

(c) $\sum\limits_{i=1}^{m} \sum\limits_{j=1}^{m} f_{ij}^2 < 1,$

then the system of m equations

$f_i(x) = x$ for $i \in \{1,2,\ldots,m\}$

has a unique solution.

7. (ii)   Consider the system of non-linear equations

$$\lambda_1 + 2\lambda_2 - 3 = 0$$
$$2\lambda_1^2 + \lambda_2^2 - 5 = 0$$

rewritten in the form

$$\lambda_1 = \left(\frac{5-\lambda_2^2}{2}\right)^{\frac{1}{2}}$$

$$\lambda_2 = \tfrac{1}{2}(3-\lambda_1)$$

on the bounded closed rectangular region

$$D \equiv \{(\lambda_1,\lambda_2) : 1 \leqslant \lambda_1 \leqslant 2, \ \tfrac{1}{2} \leqslant \lambda_2 \leqslant \tfrac{3}{2}\}.$$

Show that this system produces a matrix

$$[f_{ij}] = \begin{bmatrix} 0 & \frac{3}{\sqrt{22}} \\ \frac{1}{2} & 0 \end{bmatrix}$$

which satisfies all three conditions (a), (b) and (c).
Beginning with the initial point $\lambda_1 = \tfrac{3}{2}$, $\lambda_2 = 1$ show that
the third iteration gives approximate solutions
$\lambda_1 = 1 \cdot 478$, $\lambda_2 = 0 \cdot 755$.

8. (i)   (a) Show that the initial value problem

$$\frac{dy}{dx} = y, \quad y(0) = 1$$

is equivalent to the integral equation

$$y(x) = 1 + \int_0^x y(t)dt.$$

(b) Show that the mapping T defined by

$$Ty(x) = 1 + \int_0^x y(t)dt$$

8.   (i)   (b)   (continued)
                 is a contraction mapping of $(C[-\frac{1}{2},\frac{1}{2}],\|\cdot\|_\infty)$ into itself.

           (c)   Find the first four iterates in the Picard iteration
                 beginning from $y_0(x) = 1$ and verify that the sequence
                 of iterates converges to the solution as given by the
                 differential equation.

     (ii) (a)   Find the differential equation initial value problem
                 equivalent to the integral equation

                 $$y(x) = \int_0^x \bigl(t-y(t)\bigr)dt \ .$$

           (b)   Show that the mapping T defined by

                 $$Ty(x) = \int_0^x \bigl(t-y(t)\bigr)dt$$

                 is a contraction mapping of $(C[-\frac{1}{2},\frac{1}{2}],\|\cdot\|_\infty)$ into itself.

           (c)   Find the first four iterates in the Picard iteration
                 beginning from $y_0(x) = 0$ and verify that the sequence
                 of iterates converges to the solution as given by the
                 differential equation.

9.   (i)   Compute the first two Picard iterates for the differential
           equation

           $$y'(x) = 1 + xy^2(x) \quad \text{where } y(0) = 0.$$

     (ii) Suppose the differential equation is to hold on
           $D \equiv \{(x,y) : |x| \leqslant 1, |y| \leqslant 1\}$.
           Determine an estimate of the error in choosing the second
           iterate as a solution on the interval $[-\frac{1}{3},\frac{1}{3}]$.

10.   Given continuous real functions f and g on a bounded closed interval $[a,b]$, prove that the linear differential equation

$$y'(x) = f(x).y(x)+g(x)$$

with initial condition $y(x_0) = K$, has a unique solution in $C[a,b]$.

11.   For the following integral equations with the given initial functions determine the first three iterates and if possible determine the solution of the equations stating the range of values of the parameter λ for which the solution holds.

(i) $y(x) = 1 + \lambda \int_0^1 (x-t)y(t)dt$ with $y_0(x) = 1$.

(ii) $y(x) = 1 + \lambda \int_0^1 y(t)\sinh(x-t)dt$ with $y_0(x) = 1$.

12.   Consider a continuous function v on $[a,b]$ and a continuous function k on $D \equiv \{(x,y,z) : a \leqslant x \leqslant b, \ a \leqslant y \leqslant b\}$ which satisfies a Lipschitz condition with respect to z on D;   that is, there exists an $M > 0$ such that

$$\left|k(x,y,z_1)-k(x,y,z_2)\right| \leqslant M\left|z_1-z_2\right| \text{for all } (x,y,z_1),(x,y,z_2) \in D.$$

Prove that the non-linear integral equation

$$y(x) = v(x) + \lambda \int_a^b k\big(x,t,y(t)\big)dt$$

has one and only one solution for any $\left|\lambda\right| < \dfrac{1}{M(b-a)}$.

13. (i)   Show that the initial value problem given by the second order
          differential equation

$$\frac{d^2y}{dx^2} = f(x,y), \qquad y(x_0) = y_0, \; y'(x_0) = y_1$$

          can be transformed into a Volterra integral equation.

    (ii) (a) For the integral equation

$$y(x) = 1 + \int_0^x (x-t)y(t)\,dt$$

              determine the first three iterates in the iteration

$$y_{n+1}(x) = 1 + \int_0^x (x-t)y_n(t)\,dt \quad \text{for } n \in \mathbb{Z}^+$$

              beginning from the initial function $y_0(x) = 1$ and if
              possible determine a solution of the equation.

          (b) Transform the integral equation into a second order
               differential equation, solve it by the usual techniques
               and compare your solutions.

III. CONTINUITY

The study of continuity for real or complex functions depends mainly on the notion of distance. So there is a natural extension of the study of continuity to mappings between metric spaces. In fact the success of the generalisation of analysis to metric spaces is largely due to the fruitfulness of this general treatment of continuity.

We will see that, as in real and complex analysis continuity in metric spaces is linked closely to convergence of sequences. However, as we develop our ideas of global continuity it will become clearer that the link to convergence of sequences is not fundamental.

For spaces with linear structure, the normed linear spaces, it will be instructive to observe the interplay between the algebraic and metric structures when examining the continuity of those special mappings which preserve algebraic properties.

6. CONTINUITY IN METRIC SPACES

Our definition for continuity of a mapping between metric spaces, as with that for convergence of a sequence, is a straightforward translation of the real or complex definition with the usual metric replaced by the general metric function.

6.1 Definition. Given metric spaces (X,d) and (Y,d'), we say that a mapping $T : X \to Y$ is *continuous at* $x_0 \in X$ if given $\varepsilon > 0$ there exists a $\delta > 0$ such that

$$d'(Tx,Tx_0) < \varepsilon \quad \text{when } d(x,x_0) < \delta;$$

geometrically,

$$Tx \in B_{d'}(Tx_0; \varepsilon) \quad \text{when } x \in B_d(x_0; \delta)$$

or $\quad\quad T\left(B_d(x_0; \delta)\right) \subseteq B_{d'}(Tx_0; \varepsilon)$

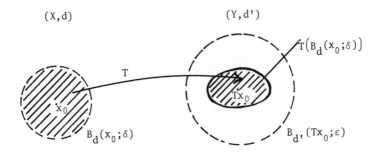

Figure 12. The continuity of the mapping T at x_0.

If T is continuous at every point of a subset A of X we say that T is *continuous on* A.

　　With continuity as with convergence of sequences, our intuition needs constant checking against the formal definitions.

6.2 Example. *Every mapping on a discrete metric space is continuous.*

Proof. Consider a mapping T from a discrete metric space (X,d) into a metric space (Y,d'). For any $x_0 \in X$, $B(x_0; 1) = \{x_0\}$ so, given $\varepsilon > 0$

$$d'(Tx, Tx_0) < \varepsilon \quad \text{when } d(x, x_0) < 1. \quad \square$$

　　Of course, the notion of continuity in such a space is practically meaningless. If all mappings are automatically continuous then there is little interest in the concept on such spaces. We noticed in Example 3.6 that discrete metric spaces possess only a degenerate form of convergence of sequences. The link is revealed in Theorem 6.3 below.

Continuity of a mapping can be characterised in terms of sequences. This serves to link the study of continuity with that of convergence of sequences as given in Section 3 and it provides a particularly useful technique for examining continuity questions.

<u>6.3 Theorem.</u> *Given metric spaces* (X,d) *and* (Y,d'), *a mapping* $T : X \to Y$ *is continuous at* $x_0 \in X$ *if and only if for every sequence* $\{x_n\}$ *in* (X,d) *convergent to* x_0, *the sequence* $\{Tx_n\}$ *in* (Y,d') *is convergent to* Tx_0.

<u>Proof.</u> Suppose that T is continuous at x_0. Then given $\varepsilon > 0$ there exists a $\delta > 0$ such that

$$d'(Tx, Tx_0) < \varepsilon \quad \text{when } d(x, x_0) < \delta.$$

Consider any sequence $\{x_n\}$ in (X,d) convergent to x_0. Then there exists a $\nu \in \mathbb{N}$ such that

$$d(x_n, x_0) < \delta \quad \text{when } n > \nu.$$

So $d'(Tx_n, Tx_0) < \varepsilon \quad \text{when } n > \nu;$

that is, the sequence $\{Tx_n\}$ in (Y,d') is convergent to Tx_0.

Conversely, suppose that T is not continuous at x_0. Then there exists an $r > 0$ such that, for each $n \in \mathbb{N}$ there exists an $x_n \in B_d(x_0; \frac{1}{n})$ where $Tx_n \notin B_{d'}(Tx_0; r)$. Then, this sequence $\{x_n\}$ is convergent to x_0 but the sequence $\{Tx_n\}$ is not convergent to Tx_0. $\square$

This sequential characterisation has an immediate implication for equivalent metrics.

<u>6.4 Corollary.</u> *Consider metric spaces* (X,d) *and* (Y,d') *and a mapping* $T : X \to Y$ *which is continuous at* $x_0 \in X$. *The mapping* T *is continuous at* x_0 *for all metrics equivalent to* d *on* X *and all metrics equivalent to* d' *on* Y; *(that is, continuity is invariant under equivalent metrics).*

The simplest way to show that a mapping T is not continuous is to find a sequence $\{x_n\}$ convergent to a point x_0 where the image sequence $\{Tx_n\}$ is not convergent to Tx_0.

<u>6.5 Example</u>. Consider the real function f on $(\mathbb{R}^2, \|\cdot\|_2)$ defined by

$$f(\lambda,\mu) = \frac{\lambda\mu}{\lambda^2+\mu^2} \,, \quad (\lambda,\mu) \neq (0,0)$$

$$f(0,0) = 0 \ .$$

Now the sequence $\left\{(\frac{1}{n},\frac{1}{n})\right\}$ converges to $(0,0)$ but $f(\frac{1}{n},\frac{1}{n}) = \frac{1}{2}$ for all $n \in \mathbb{N}$.
So f is not continuous at $(0,0)$. $\square$

<u>6.6 Example</u>. Consider the scalar function p_0 on $(C[0,1], \|\cdot\|_1)$ defined by

$$p_0(f) = f(0).$$

Now the sequence $\{f_n\}$ where

$$f_n(t) = 1 - nt \quad 0 \leqslant t \leqslant \frac{1}{n}$$

$$= 0 \qquad \frac{1}{n} < t \leqslant 1$$

satisfies $\|f_n\|_1 = \dfrac{1}{2n} \to 0$ as $n \to \infty$, so $\{f_n\}$ converges to the zero function,
but $p_0(f_n) = 1$ for all $n \in \mathbb{N}$. So p_0 is not continuous at the zero
function. $\square$

　　　　To prove that a mapping is continuous we have to decide
whether it is more expedient to aim to satisfy the definition directly or
to use the sequential approach.

<u>6.7 Example</u>. For any function $f : \mathbb{R} \to \mathbb{R}$, the induced function $h : \mathbb{R} \to \mathbb{R}^2$
which maps $\mathbb{R}$ onto G_f the graph of f, is defined by

$$h(t) = \bigl(t, f(t)\bigr).$$

Considering $\mathbb{R}$ with the usual norm and $\mathbb{R}^2$ with the Euclidean norm $\|\cdot\|_2$, we
show that if f is continuous at $t_0 \in \mathbb{R}$ then h is continuous at $t_0 \in \mathbb{R}$:

(i)   Direct method

Since f is continuous at t_0, given $\varepsilon > 0$ there exists a $0 < \delta < \varepsilon$ such that

$$\left| f(t) - f(t_0) \right| < \varepsilon \quad \text{when} \quad \left| t - t_0 \right| < \delta$$

So given $\varepsilon > 0$,

$$\left\| \left(t, f(t)\right) - \left(t_0, f(t_0)\right) \right\|_2 < \sqrt{2}\varepsilon \quad \text{when} \quad \left| t - t_0 \right| < \delta.$$

(i)'   Directly using Corollary 6.4

To reduce calculation we could consider $\mathbb{R}^2$ with the equivalent norm $\|\cdot\|_\infty$, (the product norm on $\mathbb{R}^2$, see Exercise 1.34.11). Then given $\varepsilon > 0$

$$\left\| \left(t, f(t)\right) - \left(t_0, f(t_0)\right) \right\|_\infty < \varepsilon \quad \text{when} \quad \left| t - t_0 \right| < \delta.$$

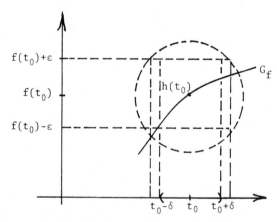

Figure 13.   Continuity of the function h at t_0 when $\mathbb{R}^2$ has norms $\|\cdot\|_2$ or $\|\cdot\|_\infty$.

(ii)   Sequential method

Since f is continuous at t_0, for every sequence $\{t_n\}$ which converges to t_0, the sequence $\{f(t_n)\}$ converges to $f(t_0)$. But then from Example 3.7, the sequence $\{\left(t_n, f(t_n)\right)\}$ converges to $\left(t_0, f(t_0)\right)$ in $(\mathbb{R}^2, \|\cdot\|_2)$. Therefore from Theorem 6.3, h is continuous at t_0. $\square$

6.8 Example. Consider the *ruler function* $f : [0,1] \to \mathbb{R}$ defined by

$$
\left.
\begin{aligned}
&f(0) = 1 \\
&f(x) = \tfrac{1}{q} \text{ for rational } x = \tfrac{p}{q} \\
&\qquad\qquad (p,q \text{ mutually prime}) \\
&f(x) = 0 \text{ for irrational } x
\end{aligned}
\right\}
$$

Now f is continuous at irrational points and discontinuous at rational points in its domain:

For any given $\varepsilon > 0$, the set $E_\varepsilon \equiv \{x \in [0,1] : f(x) \geqslant \varepsilon\}$ is finite.

Consider any irrational $x_0 \in [0,1]$. Since E_ε is finite, $\delta_0 \equiv d(x_0, E_\varepsilon) > 0$.

Then $\qquad\qquad |f(x)-f(x_0)| < \varepsilon$ for all $|x-x_0| < \delta_0$;

that is, f is continuous at irrational points.

Consider any rational $x_1 = \tfrac{p}{q} \in [0,1]$. Now $f(x_1) = \tfrac{1}{q}$. Since $E_{1/2q}$ is finite, $\delta_1 \equiv d(x_1, E_{1/2q} \backslash \{x_1\}) > 0$.

Then $\qquad\qquad |f(x)-f(x_1)| > \dfrac{1}{2q}$ for all $0 < |x-x_1| < \delta_1$;

that is, f is not continuous at rational points. □

In many cases, continuity can be deduced quite simply from an inequality.

6.9 Example. Consider $(\mathbb{R}^m, \|\cdot\|_2)$. For any $k \in \{1,2,\ldots,m\}$ the kth projection mapping $p_k : \mathbb{R}^m \to \mathbb{R}$ is defined for $x \equiv (\lambda_1, \lambda_2, \ldots, \lambda_m)$ by

$$p_k(x) = \lambda_k.$$

Now for $y \equiv (\mu_1, \mu_2, \ldots, \mu_m)$,

$$
\begin{aligned}
|p_k(x)-p_k(y)| &= |\lambda_k - \mu_k| \\
&\leqslant \sqrt{\left(\sum_{j=1}^{m} |\lambda_j - \mu_j|^2\right)} \\
&= \|x-y\|_2.
\end{aligned}
$$

It follows from this inequality that the projection mappings are continuous on $(\mathbb{R}^m, \|\cdot\|_2)$. □

6.10 Example. In any normed linear space $(X, \|\cdot\|)$, the norm $\|\cdot\|$ is a real continuous mapping on $(X, \|\cdot\|)$. This follows immediately from the inequality given in Exercise 1.34.8(ii),

$$\left| \|x\| - \|y\| \right| \leq \|x-y\| \quad \text{for all } x,y \in X. \quad \square$$

6.11 Example. A contraction mapping T on a metric space (X,d) is continuous on X. This follows immediately from Definition 5.3, for some $0 < \alpha < 1$,

$$d(Tx,Ty) \leq \alpha d(x,y) \quad \text{for all } x,y \in X. \quad \square$$

The following algebra of continuous functions theorem is a straightforward generalisation of the algebra of continuous functions theorem from real analysis.

6.12 Theorem. *Given a metric space* (X,d) *and a normed linear space* $(Y, \|\cdot\|')$ *and mappings* T_1 *and* T_2 *from X into Y, both continuous at* $x_0 \in X$, *then*   *(i)* $T_1 + T_2$ *is continuous at* x_0, *and*
            (ii) λT_1 *is continuous at* x_0 *for any scalar* λ.
When $(Y, \|\cdot\|')$ *is the scalar field* $\mathbb{R}$ *or* $\mathbb{C}$ *with the usual norm, then also*
            (iii) $T_1 . T_2$ *is continuous at* x_0, *and*
            (iv) $\dfrac{T_1}{T_2}$ *is continuous at* x_0, *provided* $T_2 x \neq 0$ *for every* $x \in X$.

6.13 Example. In Example 6.5 we considered the real function f on $(\mathbb{R}^2, \|\cdot\|_2)$ defined by

$$\left. \begin{array}{l} f(\lambda,\mu) = \dfrac{\lambda\mu}{\lambda^2+\mu^2} \quad (\lambda,\mu) \neq (0,0) \\[2mm] f(0,0) = 0 \end{array} \right\}$$

Now on $\mathbb{R}^2 \backslash \{(0,0)\}$

$$f = \frac{p_1 \cdot p_2}{\|\cdot\|_2^2}$$

where p_1 and p_2 are first and second projection mappings from $\mathbb{R}^2$ to $\mathbb{R}$ and

$\|\cdot\|_2$ is the Euclidean norm. From Examples 6.9 and 6.10, p_1, p_2 and $\|\cdot\|_2$ are continuous on $(\mathbb{R}^2, \|\cdot\|_2)$ so from Theorem 6.12, f is continuous on $\mathbb{R}^2 \setminus \{(0,0)\}$. □

Another property useful for computation is the following composition theorem.

6.14 Theorem. *Given metric spaces* (X,d), (Y,d') *and* (Z,d'') *and mappings* $T_1 : X \rightarrow Y$ *continuous at* $x_0 \in X$ *and* $T_2 : Y \rightarrow Z$ *continuous at* $T_1 x_0$, *then the composite mapping* $T_2 \circ T_1 : X \rightarrow Z$ *is continuous at* $x_0 \in X$.

Proof. Consider any sequence $\{x_n\}$ in X convergent to x_0. Since T_1 is continuous at x_0, the sequence $\{T_1 x_n\}$ is convergent to $T_1 x_0$. Since T_2 is continuous at $T_1 x_0$, the sequence $\{T_2(T_1 x_n)\}$ is convergent to $T_2(T_1 x_0)$. Therefore, from Theorem 6.3, $T_2 \circ T_1$ is continuous at $x_0 \in X$. □

If a mapping is not continuous at a point then this occurs for one of two reasons which we now explore.

6.15 Definition. Given metric spaces (X,d) and (Y,d'), a mapping $T : X \rightarrow Y$ and a cluster point x_0 of X, we say that T has a *limit as* x *approaches* x_0 *and that limit is* y_0 if given $\varepsilon > 0$ there exists a $\delta > 0$ such that

$$d'(Tx, y_0) < \varepsilon \quad \text{when} \quad 0 < d(x, x_0) < \delta;$$

geometrically,

$$Tx \in B_{d'}(y_0; \varepsilon) \quad \text{when} \quad x \in B_d(x_0; \delta) \setminus \{x_0\}$$

or $\qquad T\big(B_d(x_0; \delta) \setminus \{x_0\}\big) \subseteq B_{d'}(y_0; \varepsilon)$.

We write

$$Tx \xrightarrow[d']{} y_0 \quad \text{as} \quad x \xrightarrow[d]{} x_0$$

or $\lim\limits_{x \rightarrow x_0} Tx = y_0$ when it is obvious from the context which spaces and metrics are being used.

6.16 Remark. We notice that when we enquire into the limit behaviour of the mapping T at x_0 we examine how T maps points near x_0 but not at x_0. In the definition of limit we require x_0 to be a cluster point of X so that there will be points of the domain as close to x_0 as we wish; (see Theorem 4.7). When we consider a mapping T from a proper subset A of X, it is possible to examine the limit behaviour of T on A at a cluster point x_0 of A even if $x_0 \notin$ A and T is not defined at x_0. □

The limit behaviour of a mapping can be characterised in terms of sequences as continuity is characterised in Theorem 6.3. The proof follows a similar pattern.

6.17 Theorem. *Given metric spaces* (X,d) *and* (Y,d'), *a mapping* $T : X \to Y$ *and a cluster point* x_0 *of X, then T has a limit* y_0 *as x approaches* x_0 *if and only if for every sequence* $\{x_n\}$ *in* (X,d) *convergent to* x_0 *and* $x_n \neq x_0$, *the sequence* $\{Tx_n\}$ *in* (Y,d') *is convergent to* y_0.

Using Definition 6.15 we can say that a mapping T is continuous at a cluster point x_0 of X if and only if $\lim_{x \to x_0}$ Tx exists and is equal to Tx_0. So if T is not continuous at x_0 then either there is a failure in the limit behaviour of the mapping T at x_0 or the limit exists but is not equal to the definition of the mapping T at x_0.

6.18 Examples.

(i)   In Example 6.8 we see that at rational domain points the limit exists and is zero. Continuity fails because the limit is not equal to the definition of the function at those rational points.

(ii)   In Example 6.5 the function is discontinuous because of a failure in the limit behaviour of the function at (0,0). The sequences $\{(\frac{1}{n},\frac{1}{n})\}$ and $\{(\frac{1}{n},0)\}$ both converge to (0,0) but $f(\frac{1}{n},\frac{1}{n}) = \frac{1}{2}$ and $f(\frac{1}{n},0) = 0$ for all $n \in \mathbb{N}$. □

So far we have studied continuity as a local concept; that is, we have defined the continuity of a mapping at a point and have simply said that a continuous mapping is one which is continuous at each point of its domain. We now study continuity as a global concept and establish an important characterisation of global continuity which is of special interest in that it contains no explicit reference to metrics.

6.19 Theorem. *Given metric spaces* (X,d) *and* (Y,d'), *a mapping* $T : X \to Y$ *is continuous on* X *if and only if for every closed set* F *in* (Y,d') *the set* $T^{-1}(F)$ *is closed in* (X,d); (that is, a mapping is continuous if and only if the inverse images of closed sets are closed).

Proof. Suppose T is continuous on X and F is a closed set in (Y,d'). For any cluster point x of $T^{-1}(F)$ there exists a sequence $\{x_n\}$ in $T^{-1}(F)$ which is convergent to x. But since T is continuous at x, the sequence $\{Tx_n\}$ is convergent to Tx. Now $Tx_n \in F$ for all $n \in \mathbb{N}$ so either Tx is a member of F or is a cluster point of F. But F is closed so $Tx \in F$. Then $x \in T^{-1}(F)$; that is, $T^{-1}(F)$ is closed.

Conversely, suppose T is not continuous at x_0. Then there exists an $r > 0$ and a sequence $\{x_n\}$ in X convergent to x_0 where

$$d'(Tx_n, Tx_0) > r \quad \text{for all } n \in \mathbb{N}.$$

Consider $F \equiv \overline{\{Tx_n : n \in \mathbb{N}\}}$. Now $x_n \in T^{-1}(F)$ for all $n \in \mathbb{N}$ and since $\{x_n\}$ is convergent to x_0, then x_0 is a cluster point of $T^{-1}(F)$. But $Tx_0 \notin F$ so $x_0 \notin T^{-1}(F)$. Therefore, $T^{-1}(F)$ is not closed. $\square$

6.20 Corollary. *The kernel of a continuous mapping* T *from a metric space* (X,d) *into a normed linear space* $(Y, \|\cdot\|')$ *is a closed subset of* (X,d).

6.21 Remark. We should note that the characterisation theorem is in terms of inverse images. A continuous mapping does not necessarily map closed sets to closed sets and a mapping which does map closed sets to closed sets is not necessarily continuous. A simple example illustrating both points concerns the identity mapping between $\mathbb{R}$ with the discrete metric and $\mathbb{R}$ with the usual metric. From $\mathbb{R}$ with the discrete metric, the identity mapping is continuous but does not map closed sets to closed sets. From $\mathbb{R}$ with the usual metric, the identity mapping is not continuous but does map closed sets to closed sets. $\square$

There are some fundamental classes of mappings between metric spaces which we should distinguish.

6.22 Definition. Given metric spaces (X,d) and (Y,d'), a mapping $T : X \to Y$ is called an *isometry* if $d'(Tx,Tx') = d(x,x')$ for all $x,x' \in X$. An isometry is a one-to-one mapping which is also distance preserving.
Metric spaces (X,d) and (Y,d') are said to be *isometric* if there exists an isometry of X onto Y.

6.23 Remark. The relation, "being isometric to" is an equivalence relation on the family of all metric spaces. If two metric spaces are isometric then as metric spaces they are identical in all but name. □

6.24 Example. Given a normed linear space $(X, \|\cdot\|)$ and $x_0 \in X$, the *translation* mapping $T_{x_0} : X \to X$ defined by

$$T_{x_0}(x) = x_0 + x$$

satisfies $\quad \|T_{x_0}(x) - T_{x_0}(x')\| = \|x - x'\|$ for all $x,x' \in X$
and so is an isometry. □

As soon as we begin the study of continuous mappings from one metric space into another we become aware of the problem of determining whether the possible inverse mapping is also continuous.

6.25 Definition. Given metric spaces (X,d) and (Y,d') a mapping $T : X \to Y$ is called a *homeomorphism* if T is one-to-one and continuous and has a continuous inverse on $T(X)$.
Metric spaces (X,d) and (Y,d') are said to be *homeomorphic* if there exists a homeomorphism of X onto Y.

6.26 Remark. The relation, "being homeomorphic to" is an equivalence relation on the set of all metric spaces. It is clear that an isometry is a homeomorphism but not every homeomorphism is an isometry. So the partition of metric spaces into "homeomorphic" equivalence classes is coarser than the partition of metric spaces into "isometric" equivalence classes. □

From our sequential characterisation of continuity, Theorem 6.3, we can develop an important characterisation theorem for homeomorphisms.

6.27 Theorem. *A one-to-one mapping* T *from a metric space* (X,d) *into a metric space* (Y,d') *is a homeomorphism if and only if for any* $x_0 \in X$, *the sequence* $\{x_n\}$ *is convergent to* x_0 *in* (X,d) *if and only if the sequence* $\{Tx_n\}$ *is convergent to* Tx_0 *in* (Y,d'); (that is, a homeomorphism is a one-to-one mapping which preserves convergence of sequences).

6.28 Example. For a non-empty set X with equivalent metrics d and d', it is clear that (X,d) and (X,d') are homeomorphic under the identity mapping. But also if metric spaces (X,d) and (Y,d') are homeomorphic under the mapping T, then d and d* where

$$d^*(x,x') = d'(Tx,Tx') \quad \text{for all } x,x' \in X$$

are equivalent metrics for X. $\square$

From the inverse image characterisation of continuity given in Theorem 6.19 we can develop another useful characterisation theorem for homeomorphisms.

6.29 Theorem. *A one-to-one mapping* T *from a metric space* (X,d) *onto a metric space* (Y,d') *is a homeomorphism if and only if for every closed set* E *in* (X,d) *the set* $T(E)$ *is closed in* (Y,d') *and for every closed set* F *in* (Y,d') *the set* $T^{-1}(F)$ *is closed in* (X,d); (that is, a homeomorphism onto is a one-to-one mapping which preserves closed sets).

Proof. From Theorem 6.19, T^{-1} is continuous on (Y,d') if and only if for every closed set E in (X,d), $(T^{-1})^{-1}(E)$ is closed in (Y,d'). But since T is one-to-one, $(T^{-1})^{-1}(E) = T(E)$. $\square$

It is often convenient to use this result as a means of deducing that certain sets are closed.

6.30 Example. Given a normed linear space $(X, \| \cdot \|)$ with proper closed
linear subspace M , for any $x_0 \in X$, the coset $x_0 + M$ is closed:
The coset $x_0 + M$ is the image of M under the translation map T_{x_0} given in
Example 6.24. But T_{x_0} is a homeomorphism of X onto X, so from Theorem 6.29
we deduce that $x_0 + M$ is closed. ◻

In real analysis, uniform continuity is an important type of
global continuity. Real uniformly continuous functions can be approximated by step functions and this is the key to proving the Riemann
integrability of continuous functions. The generalisation of the notion
of uniform continuity to metric spaces is quite straightforward and is
useful in the more abstract setting.

6.31 Definition. Given metric spaces (X,d) and (Y,d') a mapping $T : X \to Y$
is said to be *uniformly continuous* on X if given $\varepsilon > 0$ there exists a
$\delta > 0$ such that for all $x,x' \in X$,

$$d'(Tx,Tx') < \varepsilon \quad \text{when } d(x,x') < \delta.$$

6.32 Remark. It is clear that uniform continuity implies continuity but
global continuity does not necessarily imply uniform continuity. For
continuity at $x \in X$, given $\varepsilon > 0$ there exists a $\delta > 0$ which may depend on
both ε and x but for uniform continuity on X, given $\varepsilon > 0$ there exists a
$\delta > 0$ which depends on ε but is independent of any particular $x \in X$. ◻

All mappings where the continuity of the mapping can be
deduced from a global inequality are uniformly continuous.

6.33 Example. A real function f which has bounded derivative on an
interval J is uniformly continuous on J:
Now there exists an $M > 0$ such that $|f'(t)| \leqslant M$ for all $t \in J$. So we
deduce from the Mean Value Theorem for differential calculus that f
satisfies a Lipschitz condition

$$|f(s)-f(t)| \leqslant M|s-t| \quad \text{for all } s,t \in J$$

and this implies that f is uniformly continuous on J.

The function f on $(0,\infty)$ defined by

$$f(t) = \frac{1}{t}$$

is not uniformly continuous on $(0,\infty)$. But it is possible to have a uniformly continuous differentiable function with an unbounded derivative. For example, the function f on $(0,\infty)$ defined by

$$f(t) = \sqrt{t}$$

is uniformly continuous on $(0,\infty)$. $\square$

We notice that Examples 6.9, 6.10 and 6.11, where continuity was deduced from an inequality, are all examples of uniformly continuous mappings.

6.34 Example. For any metric spaces (X,d) and (Y,d') an isometry T from X into Y satisfies

$$d'(Tx,Tx') = d(x,x') \quad \text{for all } x,x' \in X$$

and so is uniformly continuous on X. $\square$

We noticed in Theorem 6.3 that a continuous mapping maps convergent sequences to convergent sequences. However, continuous mappings do not necessarily map Cauchy sequences to Cauchy sequences.

6.35 Example. Consider the real function tan on $(-\frac{\pi}{2},\frac{\pi}{2})$. The sequence $\{x_n\}$ where $x_n = \tan^{-1} n$ is a Cauchy sequence in $(-\frac{\pi}{2},\frac{\pi}{2})$ but the sequence $\{\tan x_n\}$ has the property that

$$\left| \tan x_n - \tan x_m \right| = \left| n - m \right|$$

and is not Cauchy. $\square$

However, uniformly continuous mappings can be characterised by their mapping of Cauchy sequences to Cauchy sequences.

6.36 Theorem. *Given metric spaces* (X,d) *and* (Y,d'), *a mapping* $T : X \to Y$ *is uniformly continuous on* X *if and only if for every Cauchy sequence* $\{x_n\}$ *in* (X,d), *the sequence* $\{Tx_n\}$ *is Cauchy in* (Y,d').

Proof. If T is uniformly continuous on X then, given $\varepsilon > 0$ there exists a $\delta > 0$ such that for all $x, x' \in X$,

$$d'(Tx,Tx') < \varepsilon \quad \text{when } d(x,x') < \delta.$$

If $\{x_n\}$ is Cauchy in (X,d) there exists a $\nu \in \mathbb{N}$ such that

$$d(x_n, x_m) < \delta \quad \text{for all } m,n > \nu.$$

But then

$$d'(Tx_n, Tx_m) < \varepsilon \quad \text{for all } m,n > \nu;$$

that is, $\{Tx_n\}$ is Cauchy in (Y,d').

Conversely, if T is not uniformly continuous on X, then there exists an $r > 0$ and a sequence $\{x_n\}$ in (X,d) such that

$$d(x_n, x_m) < \tfrac{1}{n} \quad \text{for all } m > n$$

but $\qquad d'(Tx_n, Tx_m) > r \quad \text{for all } m > n.$

Then $\{x_n\}$ is a Cauchy sequence in (X,d) but $\{Tx_n\}$ is not a Cauchy sequence in (Y,d'). $\square$

Such a sequential characterisation for uniform continuity is useful in providing a simple way to demonstrate non uniform continuity as we have done in Example 6.35.

Theorem 6.36 enables us to establish a useful extension property for uniformly continuous mappings.

6.37 Theorem. *Given a metric space* (X,d) *with a dense subset* A *in* X *and a complete metric space* (Y,d'), *a uniformly continuous mapping* $T : A \to Y$ *has a unique continuous extension* $\tilde{T}$ *from* X *into* Y *and* $\tilde{T}$ *is uniformly continuous on* X.

Proof. Consider $x \in C(A)$. Since x is a cluster point of A in (X,d) there exists a sequence $\{a_n\}$ in A convergent to x. But since T is uniformly continuous on A we have from Theorem 6.36 that $\{Ta_n\}$ is a Cauchy sequence in (Y,d'). But (Y,d') is complete so there exists a $y \in Y$ such that $\{Ta_n\}$ converges to y. We define an extension $\tilde{T}$ on X by $\tilde{T}(x) = y$.

We need to show that this extension is well defined; that is, that the definition of $\tilde{T}$ at x is independent of the particular sequence $\{a_n\}$ chosen to converge to x:

For a sequence $\{a_n'\}$ in A converging to x we have the sequence $\{Ta_n'\}$ converging to y'. Consider the sequence $\{a_1, a_1', a_2, a_2', \ldots, a_n, a_n', \ldots\}$ converging to x. Now from Theorem 6.36 we have that the image of this sequence under T is Cauchy. Therefore $y' = y$.

Now from Theorem 6.3 we see that $\tilde{T}$ as defined is continuous at x. We need to show that $\tilde{T}$ is uniformly continuous on X:

We have that T is uniformly continuous on A; that is, given $\varepsilon > 0$ there exists a $\delta > 0$ such that for all $a, a' \in A$, $d'(Ta, Ta') < \varepsilon$ when $d(a, a') < \delta$. Consider $x, x' \in X$ such that $d(x, x') < \frac{\delta}{2}$. From the definition of $\tilde{T}$ there exists an $a \in A$ such that $d'(\tilde{T}x, \tilde{T}a) < \varepsilon$ and $d(x, a) < \frac{\delta}{4}$ and an $a' \in A$ such that $d'(Tx', Ta') < \varepsilon$ and $d(x', a') < \frac{\delta}{4}$. Therefore

$$d(a, a') \leqslant d(a, x) + d(x, x') + d(a', x')$$
$$< \delta$$

and since T is uniformly continuous on A, we have $d'(Ta, Ta') < \varepsilon$. Therefore

$$d'(\tilde{T}x, \tilde{T}x') \leqslant d'(\tilde{T}x, \tilde{T}a) + d'(\tilde{T}a, \tilde{T}a') + d'(\tilde{T}a', \tilde{T}x')$$
$$< 3\varepsilon$$

and we conclude that $\tilde{T}$ is uniformly continuous on X. $\square$

This theorem can be applied to extending continuous real functions from the rationals $\mathbb{Q}$ to the reals $\mathbb{R}$.

6.38 Example. Consider the exponential function f defined on $\mathbb{Q}$ for any given $a > 1$, by

$$f(r) = a^r.$$

Now for $r \in \mathbb{Q}$, $r > 0$ we have $1 - ar < a^{-r}$ so for $r, r' \in Q$ where $r' > r$,

$$0 < a^{r'} - a^r = a^{r'}\left((1 - a^{-(r'-r)})\right)$$
$$< a^{r'+1} \cdot (r'-r).$$

Therefore f is uniformly continuous on any bounded rational interval J. So we can define the exponential function $\tilde{f}$ on $\mathbb{R}$ as the unique continuous extension of f. By $a^{\sqrt{2}}$ we mean $\lim\limits_{n \to \infty} a^{r_n}$ where $\{r_n\}$ is a sequence of rationals converging to $\sqrt{2}$. $\square$

6.39 Exercises.

1.   For the following real functions examine the limit behaviour at the points specified.

   (i)   On $\mathbb{R}$ with the usual norm,

      (a)   f defined on $\mathbb{R} \setminus \{2\}$ where

$$f(t) = \frac{t^2 - 4}{t - 2}, \text{ at } t = 2.$$

      (b)   f defined on $\mathbb{R} \setminus \{0\}$ where

$$f(t) = \sin \frac{1}{t}, \text{ at } t = 0.$$

      (c)   f defined on $\mathbb{R}$ where

$$f(t) = t \quad t \text{ rational}$$
$$= 1-t \quad t \text{ irrational} \Big\},$$

at $t = 0$ and $t = \frac{1}{2}$.

      (d)   f defined on $\mathbb{Q} \cap (0,1)$ where

$$f(x) = \frac{1}{q} \text{ for } x = \frac{p}{q}, \quad p, q \text{ mutually prime,}$$

at $x = 0$, $\frac{1}{2}$ and $\frac{\sqrt{2}}{2}$.

1. (ii)   On $(\mathbb{R}^2, \|\cdot\|_2)$ for real functions defined on $\mathbb{R}^2 \setminus \{(0,0)\}$, at $(\lambda,\mu) = (0,0)$,

(a) $f(\lambda,\mu) = \dfrac{\lambda^2 - \mu^2}{\lambda^2 + \mu^2}$,

(b) $f(\lambda,\mu) = \dfrac{\lambda\mu}{|\lambda| + |\mu|}$,

(c) $f(\lambda,\mu) = \dfrac{\lambda\mu^2}{\lambda^2 + \mu^4}$,

(d) $f(\lambda,\mu) = \dfrac{\sin(|\lambda| + |\mu|)}{\sqrt{\lambda^2 + \mu^2}}$.

2. (i)   Consider $(\mathbb{R}^n, \|\cdot\|)$ and $(\mathbb{R}^m, \|\cdot\|')$ and a mapping T from $\mathbb{R}^n$ into $\mathbb{R}^m$. Prove that T is continuous at $x_0 \in \mathbb{R}^n$ if and only if $p_k \circ T : \mathbb{R}^n \to \mathbb{R}$ is continuous at x_0 for each $k \in \{1,2,\ldots,m\}$ where p_k is the kth projection mapping from $\mathbb{R}^m$ into $\mathbb{R}$ defined for $y \equiv \{\mu_1, \mu_2, \ldots, \mu_m\}$ by

$$p_k(y) = \mu_k.$$

(ii)   Deduce that the following mappings are continuous under any norms on the domain and range spaces

(a) $T : \mathbb{R}^2 \to \mathbb{R}^3$ where

$$T(\lambda,\mu) = (\lambda, \lambda+\mu, \lambda-\mu),$$

(b) $T : \mathbb{R}^3 \to \mathbb{R}^2$ where

$$T(\lambda,\mu,\nu) = (\lambda+\mu+\nu, \lambda\mu\nu),$$

(c) $T : \mathbb{R}^2 \to \mathbb{R}^2$ where

$$T(\lambda,\mu) = (\max\{|\lambda|, |\mu|\}, \min\{|\lambda|, |\mu|\}).$$

3.   (i)   Given normed linear spaces $(X, \|\cdot\|)$ and $(Y, \|\cdot\|')$ prove that if
the mapping $T : X \to Y$ is continuous at $x_0 \in X$ and a real
mapping g on X is continuous at x_0 then the product mapping
$g.T : X \to Y$ is continuous at x_0.

      (ii)   Given a normed linear space $(X, \|\cdot\|)$ prove that the operator T
on X defined by

$$T(x) = \frac{x}{\|x\|} \quad x \neq 0$$
$$T(0) = 0$$

is continuous on $X \backslash \{0\}$ but is not continuous at 0.

4.   (i)   Given that the real function f on $(\mathbb{R}^2, \|\cdot\|_2)$ is continuous at
(λ_0, μ_0), prove that the real function f_{μ_0} on $\mathbb{R}$ with the
usual norm, defined by

$$f_{\mu_0}(\lambda) = f(\lambda, \mu_0)$$

is continuous at λ_0.

      (ii)   Given a real function f on $\mathbb{R}$ with the usual norm, and the
induced function h which maps $\mathbb{R}$ onto G_f in $(\mathbb{R}^2, \|\cdot\|_2)$ defined by

$$h(t) = \big(t, f(t)\big),$$

prove that if h is continuous at t_0 then f is continuous at
t_0.

5.   *Cantor's mapping* f maps $[0,1]$ in ternary representation into
$[0,1]$ in binary representation and is defined

for $x \equiv \sum_{k=1}^{\infty} \frac{a_k}{3^k} \equiv \cdot a_1 a_2 \ldots a_n \ldots, \quad a_k \in \{0,1,2\}, \quad$ (ternary)

by $f(x) = \sum_{k=1}^{\infty} \frac{b_k}{2^k} \equiv \cdot b_1 b_2 \ldots b_n \ldots, b_k \in \{0,1\}, \quad$ (binary)

where, if the ternary representation has digits $a_k \in \{0,2\}$ for all $k \in \mathbb{N}$
we put $b_k = \frac{1}{2}a_k$ and if the ternary representation has some digit $a_k = 1$

and ν is the smallest $k \in \mathbb{N}$ such that $a_k = 1$ we put

$$
\begin{array}{ll}
b_k = \tfrac{1}{2}a_k & k < \nu \\
b_\nu = 1 & \\
b_k = 0 & k > \nu
\end{array}
\left.\vphantom{\begin{array}{l}a\\b\\c\end{array}}\right\}
$$

Prove that   (i)   f maps Cantor's Ternary set T onto $[0,1]$.

(ii)   f is increasing and

(iii)   f is continuous;

(see Exercise 4.51.6).

6.   (i)   Given metric spaces (X,d) and (Y,d') and continuous mappings S and T from X into Y, prove that the set $\{x \in X : Sx = Tx\}$ is closed in (X,d).

(ii)   Given an alternative proof in the case where S and T are continuous mappings from a metric space (X,d) into a normed linear space $(Y, \|\cdot\|')$.

7.   (i)   Consider the normed linear space $(m, \|\cdot\|_\infty)$ and the functional f on m where for $x \equiv \{\lambda_1, \lambda_2, \ldots, \lambda_n, \ldots\}$,

$$f(x) = \lim \sup\{\lambda_n : n \in \mathbb{N}\}.$$

(a)   Prove that f is continuous on $(m, \|\cdot\|_\infty)$.

(b)   Deduce that the subset

$$\{x \in m : 0 \leq \lim \sup\{\lambda_n : n \in \mathbb{N}\} \leq 1\}$$

is a closed subset of $(m, \|\cdot\|_\infty)$.

7. (ii)   Consider the normed linear space $(C[-1,1], \|\cdot\|_\infty)$ and the
             functional F on $C[-1,1]$ where

$$F(f) = \frac{f(t)+f(-t)}{2} \quad \text{for } t \in [-1,1].$$

(a) Prove that F is continuous on $(C[-1,1], \|\cdot\|_\infty)$

(b) Deduce that the subset

$$\{f \in C[-1,1] : F(f) = f\},$$

(the subset of *even functions*), is a closed subset of
$(C[-1,1], \|\cdot\|_\infty)$.

(iii)   Consider the normed linear spaces $(c_0, \|\cdot\|_\infty)$ and $(E_0, \|\cdot\|_\infty)$.
         The mapping $T : c_0 \to E_0$ is defined by

$$\{\lambda_1, \lambda_2, \ldots, \lambda_n, \ldots\} \longmapsto \{\mu_1, \mu_2, \ldots, \mu_n, \ldots\}$$

where $\mu_k = \lambda_k$ if $|\lambda_k| \geq 1$
             $= 0$ if $|\lambda_k| < 1$

(a) Prove that T is continuous on $(c_0, \|\cdot\|_\infty)$.

(b) Deduce that the subset

$$\{x \in E_0 : |\lambda_n| \geq 1 \text{ for all non-zero elements } \lambda_n\}$$

is a closed subset of $(c_0, \|\cdot\|_\infty)$.

8.   A real function ϕ defined on a convex subset K of a linear
space X is said to be a *convex function* if

$$\phi(\lambda x+(1-\lambda)y) \leq \lambda\phi(x) + (1-\lambda)\phi(y) \text{ for all } x,y \in K \text{ and } 0 \leq \lambda \leq 1.$$

(i) For $K \equiv [a,b]$ an interval in $\mathbb{R}$ with the usual norm, prove that
     ϕ is continuous on K.

(ii) For $K \equiv \{(\lambda, \mu) \in \mathbb{R}^2 : \max\{|\lambda|, |\mu|\} \leq 1\}$ in $(\mathbb{R}^2, \|\cdot\|_2)$, prove that
     ϕ is continuous on K.

9. (i)   (a) For a metric space (X,d), given $x_0 \in X$, prove that the
                real function f on X defined by

$$f(x) = d(x,x_0)$$

                is uniformly continuous on X.

         (b) For a metric space (X,d), given a subset A of X, prove
                that the real function ϕ on X defined by

$$\phi(x) = d(x,A)$$

                is uniformly continuous on X;
                (see Exercise 2.33.6(i)).

    (ii)   (a) For a normed linear space $(X, \|\cdot\|)$, prove that the
                function f defined in (i)(a) is a convex function.

         (b) For a normed linear space $(X, \|\cdot\|)$ where A is a convex
                set, prove that the function ϕ defined in (i)(b) is a
                convex function.

10.   Consider metric spaces (X,d) and (Y,d') and the product space
$(X \times Y, d_\pi)$, (see Exercise 1.34.11(i))

    (i)   Prove that a sequence $\{(x_n,y_n)\}$ converges to (x_0,y_0) in
           $(X \times Y, d_\pi)$ if and only if $\{x_n\}$ converges to x_0 in (X,d) and $\{y_n\}$
           converges to y in (Y,d').

    (ii)   A mapping $T : X \to Y$ generates a graph $G_T \equiv \{(x,Tx) : x \in X\}$ in
           $X \times Y$. Prove that G_T is closed in $(X \times Y, d_\pi)$ if and only if for
           every sequence $\{x_n\}$ converging to x in (X,d) and sequence $\{Tx_n\}$
           converging to y in (Y,d') we have $y = Tx$.

    (iii) Prove that if the mapping T is continuous on X then the graph
           G_T is closed in $(X \times Y, d_\pi)$. Give an example from real analysis
           to show that the converse is not true in general.

11.   Consider metric spaces (X,d) and (Y,d') and the product space $(X \times Y, d_\pi)$, (See exercise 1.34.11(i)).

    (i)   Prove that the projection mappings

$$p_1 : X \times Y \to X \quad \text{where} \quad p_1(x,y) = x \quad \text{and}$$

$$p_2 : X \times Y \to Y \quad \text{where} \quad p_2(x,y) = y$$

are continuous on $X \times Y$.

    (ii)   Given a mapping T from a metric space (Z,d'') into $(X \times Y, d_\pi)$, prove that T is continuous on Z if and only if both $p_1 \circ T : Z \to X$ and $p_2 \circ T : Z \to Y$ are continuous.

12. (i)   Given metric spaces (X,d) and (Y,d') and a mapping $T : X \to Y$, prove that T is continuous on X if and only if for every subset A of X we have $T(\overline{A}) \subseteq \overline{T(A)}$.

    (ii)   Prove that if T is one-to-one and onto then T is a homeomorphism if and only if $T(\overline{A}) = \overline{T(A)}$.

13. (i)   Prove that a real continuous periodic function f on $\mathbb{R}$ is uniformly continuous, (f is *periodic* if for some $a \in \mathbb{R}$, $f(t+a) = f(t)$ for all $t \in \mathbb{R}$).

    (ii)   Given a real function f uniformly continuous on $\mathbb{R}$, prove that the sequence $\{f_n\}$ where

$$f_n(t) = f(t + \tfrac{1}{n})$$

is uniformly convergent in $C[0,1]$.

14. (i)   (a) Prove that a real uniformly continuous function on a bounded interval is bounded.
        (b) Prove that an unbounded real function on a bounded interval cannot be uniformly continuous.

14. (ii)   (a)   Prove that any mapping T from a discrete metric space
                  (X,d) into a metric space (Y,d') is uniformly continuous.

            (b)   Give an example to show that for mappings between metric
                  spaces in general, the image of a bounded set under a
                  uniformly continuous mapping is not necessarily bounded.

15. (i)   Given metric spaces (X,d), (Y,d') and (Z,d") and mappings
            $T : X \to Y$ and $S : Y \to Z$, prove that if T and S are uniformly
            continuous then $S \circ T : X \to Z$ is uniformly continuous.

     (ii)   Given metric spaces (X,d) and (Y,d') and a homeomorphism T
            from X onto Y, if T is uniformly continuous is T^{-1} necessarily
            uniformly continuous too? Prove or give a counter-example.

16.   Consider a normed linear space $(X, \|\cdot\|)$ and a real functional f
on X with the property that

$$f(x+y) = f(x)f(y) \quad \text{for all } x,y \in X.$$

     (i)   Prove that f is continuous on X if and only if f is continuous
            at $\underset{\sim}{0}$.

     (ii)   Prove that if f is continuous at $\underset{\sim}{0}$ then f is uniformly
            continuous on any bounded subset of X.

7. CONTINUOUS LINEAR MAPPINGS

The natural mappings between normed linear spaces which deserve special study are the linear mappings; they are the homomorphisms which preserve the linear structure. We would expect that the algebraic character of these special mappings would bring certain simplifications to the study of their continuity.

As a first observation we note that there is no distinction to be drawn between the study of their local and global continuity.

7.1 Theorem. *Given normed linear spaces* $(X, \|\cdot\|)$ *and* $(Y, \|\cdot\|')$, *a linear mapping* $T : X \to Y$ *is continuous on* X *if and only if* T *is continuous at* $\underset{\sim}{0}$.

Proof. We need only show that local continuity at $\underset{\sim}{0}$ implies global continuity on X. Since T is linear, $T(\underset{\sim}{0}) = \underset{\sim}{0}$. If T is continuous at $\underset{\sim}{0}$ then given $\varepsilon > 0$ there exists a $\delta > 0$ such that

$$\|Tx\|' < \varepsilon \quad \text{when} \quad \|x\| < \delta.$$

But this implies that

$$\|T(x-y)\|' < \varepsilon \quad \text{when} \quad \|x-y\| < \delta.$$

Since T is linear this implies that

$$\|Tx-Ty\|' < \varepsilon \quad \text{when} \quad \|x-y\| < \delta.$$

We conclude that T is uniformly continuous on X. □

In normed linear spaces it is natural to consider the behaviour of a linear mapping at $\underset{\sim}{0}$, but a more general statement is clearly true.

7.2 Corollary. *Given normed linear spaces* $(X, \|\cdot\|)$ *and* $(Y, \|\cdot\|')$, *a linear mapping* $T : X \to Y$ *is continuous on* X *if and only if* T *is continuous at any particular point* $x_0 \in X$.

7.3 Remark. It follows from Corollary 7.2 that a linear mapping is either continuous at every point or discontinuous at every point. □

7.4 Example. For the projection mapping $p_k : \mathbb{R}^m \to \mathbb{R}$ discussed in Example 6.9 and defined for $x \equiv (\lambda_1, \lambda_2, \ldots, \lambda_m)$ by

$$p_k(x) = \lambda_k,$$

we notice that p_k is a linear functional on $\mathbb{R}^m$ so the continuity of p_k on $\mathbb{R}^m$ is determined by the continuity of p_k at $\underline{0}$. Now the inequality

$$|p_k(x)| = |\lambda_k| \leq \sqrt{(\sum_{j=1}^{m} |\lambda_j|^2)} = \|x\|_2$$

implies that p_k is continuous at $\underline{0}$. $\square$

7.5 Example. Consider $(R[a,b], \|\cdot\|_\infty)$ and the linear functional I on $R[a,b]$ defined by

$$I(f) = \int_a^b f(t)dt$$

Now
$$|I(f)| = |\int_a^b f(t)dt|$$
$$\leq \int_a^b |f(t)|dt \leq (b-a)\|f\|_\infty.$$

So I is continuous at $\underline{0}$ and consequently is continuous on $(R[a,b], \|\cdot\|_\infty)$. $\square$

7.6 Remark. This is quite a significant continuous linear functional and should be considered in relation to Example 4.23 where it was shown that the normed linear space $(R[a,b], \|\cdot\|_\infty)$ is complete. One of the classical uniform convergence theorems reads as follows.

The limit of a uniformly convergent sequence $\{f_n\}$ *of Riemann integrable functions on* $[a,b]$ *is*

    (i)   Riemann integrable and

    (ii) $\lim\limits_{n \to \infty} \int_a^b f_n(t)dt = \int_a^b \lim\limits_{n \to \infty} f_n(t)dt.$

Now (i) is guaranteed by Example 4.23 and (ii) is saying that the integral is a continuous functional with respect to the uniform norm and this is proved in Example 7.5. $\square$

However, not all linear mappings are continuous.

7.7 Example. In Example 6.6 we discussed the real mapping p_0 on $(C[0,1], \|\cdot\|_1)$ defined by

$$p_0(f) = f(0).$$

Now p_0 is a linear functional on $C[0,1]$. We saw by examining a sequence $\{f_n\}$ converging to the zero function that p_0 is not continuous at the zero function so p_0 is not continuous on $(C[0,1], \|\cdot\|_1)$. $\square$

7.8 Example. Consider $(C^1[-\pi,\pi], \|\cdot\|_\infty)$ the linear space of continuously differentiable functions on $[-\pi,\pi]$ with the uniform norm and the linear functional D_0 on $C^1[-\pi,\pi]$ defined by

$$D_0(f) = f'(0).$$

Consider the sequence $\{f_n\}$ in $C^1[-\pi,\pi]$

where $\qquad f_n(t) = \frac{1}{n}\sin nt.$

Now $\qquad \|f_n\|_\infty = \frac{1}{n} \to 0 \quad \text{as} \quad n \to \infty,$

so $\{f_n\}$ is uniformly convergent to the zero function.

But $\qquad D_0(f_n) = f_n'(0) = 1 \quad \text{for all } n \in \mathbb{N}$

$$\not\to 0 \quad \text{as } n \to \infty.$$

So D_0 is not continuous at the zero function and is not continuous on $(C^1[-\pi,\pi], \|\cdot\|_\infty)$. $\square$

7.9 Remark. This is a significant discontinuous linear functional and should be considered in relation to Example 4.24 where it was shown that the normed linear space $(C^1[-\pi,\pi], \|\cdot\|_\infty)$ is not complete. It is not possible to establish a classical uniform convergence theorem for differentiation along the lines of that for integration given in Remark 7.6. In Example 4.24 we examined a sequence $\{f_n\}$ which was uniformly convergent to a non-differentiable function, so there is no equivalent to (i) in

Remark 7.6 for differentiation. By proving that D_0 is a discontinuous linear functional in Example 7.8 we have shown that even if a uniformly convergent sequence has a limit which is differentiable, still it is not true in general that

$$\lim_{n\to\infty} \frac{d}{dt}\big(f_n(t)\big) = \frac{d}{dt}\big(\lim_{n\to\infty} f_n(t)\big)$$

and so there is no equivalent to (ii) in Remark 7.6 for differentiation. □

        Nevertheless, intuitively we might tend to assume that in many cases, the algebraic structure of a linear mapping would automatically imply continuity. This is in fact the case for linear mappings defined on finite dimensional normed linear spaces.

7.10 Theorem. *A linear mapping from a finite dimensional normed linear space into a normed linear space is continuous.*

Proof. Consider an m-dimensional normed linear space $(X_m, \|\cdot\|)$ where X_m has a basis $\{e_1, e_2, \ldots, e_m\}$. Now for $x \equiv \lambda_1 e_1 + \lambda_2 e_2 + \ldots + \lambda_m e_m$ the co-ordinate functionals f_k where $k \in \{1, 2, \ldots, m\}$ defined by

$$f_k(x) = \lambda_k$$

are linear functionals on X_m and Theorem 3.8 implies that each f_k is continuous.

For any linear mapping T from $(X_m, \|\cdot\|)$ into a normed linear space $(Y, \|\cdot\|')$ we have

$$Tx = \lambda_1 Te_1 + \lambda_2 Te_2 + \ldots + \lambda_m Te_m$$
$$= (Te_1 f_1 + Te_2 f_2 + \ldots + Te_m f_m)(x) \quad \text{for all } x \in X_m.$$

But for each $k \in \{1, 2, \ldots, m\}$, the mapping $Te_k f_k : X_m \to sp\{Te_k\}$ defined by

$$Te_k f_k(x) = \lambda_k Te_k$$

is linear and continuous.

So $T = Te_1 f_1 + Te_2 f_2 + \ldots + Te_m f_m$ as a finite sum of continuous mappings is by Theorem 6.12 also continuous.   □

We should draw attention to the following particular case.

7.11 Corollary. *All linear functionals on a finite dimensional normed linear space are continuous.*

7.12 Remark. Notice that Theorem 7.10 enables us to deduce without any computation except for an algebraic check on linearity, that the projection mappings p_k of Example 7.4 are continuous under any norm on $\mathbb{R}^m$. Theorem 7.10 provides the basic reason why the study of linear mappings on finite dimensional normed linear spaces is essentially confined to linear algebra and in particular to matrix theory. It is only when we include the study of infinite dimensional normed linear spaces that the continuity of the linear mappings assumes interest. □

It was mentioned in Section 6 that in some cases continuity can be conveniently deduced from an inequality. The continuity of linear mappings can always be established in this way.

7.13 Theorem. *Given normed linear spaces $(X, \|\cdot\|)$ and $(Y, \|\cdot\|')$ a linear mapping $T : X \to Y$ is continuous if and only if there exists an $M > 0$ such that*

$$\|Tx\|' \leq M\|x\| \quad \text{for all } x \in X.$$

Proof. Suppose that there exists an $M > 0$ such that $\|Tx\|' \leq M\|x\|$ for all $x \in X$. Then clearly T is continuous at $\underline{0}$.

Conversely, suppose that the condition does not hold; that is, for each $n \in \mathbb{N}$ there exists an $x_n \in X$ such that

$$\|Tx_n\|' > n\|x_n\|.$$

Then $\left\| T\left(\frac{1}{n} \frac{x_n}{\|x_n\|}\right) \right\|' > 1$ for all $n \in \mathbb{N}$.

But sequence $\{\frac{1}{n} \frac{x_n}{\|x_n\|}\}$ converges to $\underset{\sim}{u}$ and $\{T(\frac{1}{n} \frac{x_n}{\|x_n\|})\}$ does not converge to $\underset{\sim}{0}$ and this implies that T is not continuous at $\underset{\sim}{0}$. $\square$

7.14 Remark. Given normed linear spaces $(X, \|\cdot\|)$ and $(Y, \|\cdot\|')$, a linear mapping $T : X \to Y$ where there exists an $M > 0$ such that

$$\|Tx\|' \leq M\|x\| \quad \text{for all } x \in X$$

is sometimes said to be a *bounded* linear mapping. The reason for this terminology is that T satisfies such an inequality if and only if T maps bounded sets in $(X, \|\cdot\|)$ into bounded sets in $(Y, \|\cdot\|')$. However, since Theorem 7.13 tells us that T is bounded if and only if T is continuous there is little point to this terminology in this present context. $\square$

7.15 Remark. To prove the continuity of a linear mapping we should aim to establish the boundedness inequality of Theorem 7.13; this was the procedure followed in Examples 7.4 and 7.5. To show non-continuity of a linear mapping we should try to find a sequence converging to $\underset{\sim}{0}$ which does not conform to the sequential continuity condition; this was the method followed in Examples 7.7 and 7.8. $\square$

Corollary 6.20 is particularly useful in the study of continuous linear mappings. Given normed linear spaces $(X, \|\cdot\|)$ and $(Y, \|\cdot\|')$ and a linear mapping $T : X \to Y$ then ker T is a linear subspace, so if T is continuous then ker T is a closed linear subspace.

For linear functionals this property of the kernel being closed characterises continuity.

7.16 Theorem. *Given a normed linear subspace* $(X, \|\cdot\|)$, *a linear functional f on X is continuous if and only if* ker f *is closed.*

Proof. We need only prove that if ker f is closed then f is continuous. Suppose that ker f is closed. If f is the zero functional then f is continuous. If f is not the zero functional then there exists an $x_0 \notin$ ker f such that $f(x_0) = 1$. Since ker f is closed we have from Example 6.30 that the coset $x_0 +$ ker f is also closed. Since $\underset{\sim}{0} \notin x_0 +$ ker f then from Theorem 4.15 there exists an $r > 0$ such that $B(\underset{\sim}{0}; r) \cap (x_0 + \text{ker } f) = \phi$.

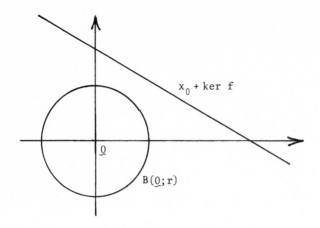

Figure 14. 0 is not a cluster point of the closed
coset $x_0 + \ker f$.

We show that $|f(x)| \leq \frac{1}{r}\|x\|$ for all $x \in X$:

Suppose on the contrary, that there exists some $x_1 \in X$ such that
$|f(x_1)| > \frac{1}{r}\|x_1\|$.

Then $\frac{x_1}{f(x_1)} \in B(\underline{0};r)$.   However, $\frac{x_1}{f(x_1)} - x_0 \in \ker f$;

that is, $\frac{x_1}{f(x_1)} \in x_0 + \ker f$.   But this contradicts the fact that

$B(\underline{0};r) \cap (x_0 + \ker f) = \phi$. $\Box$

7.17 Remark.   This theorem does not extend to linear mappings generally:
Consider the identity mapping id from $(C[0,1], \|\cdot\|_1)$ onto $(C[0,1], \|\cdot\|_\infty)$.
Now $\ker \text{id} = \{\underline{0}\}$ which is closed.   But as in Example 3.17 the sequence
$\{f_n\}$ where

$$f_n(t) = t^n$$

satisfies $\|f_n\|_1 \to 0$ as $n \to \infty$, but $\|f_n\|_\infty = 1$ for all $n \in \mathbb{N}$ so id is not
continuous. $\Box$

It is worthwhile exploring further the nature of the kernel
of a linear functional on a normed linear space.

7.18 Lemma. *If f is a non-zero linear functional on a linear space X and $x_0 \notin$ ker f then each $y \in X$ can be represented uniquely in the form*

$$y = \lambda x_0 + z \quad where \quad z \in ker\ f.$$

Proof. Since $x_0 \notin$ ker f then $f(x_0) \neq 0$. If we put $\lambda \equiv \dfrac{f(y)}{f(x_0)}$ and define $z \equiv y - \dfrac{f(y)}{f(x_0)} x_0$ then $y = \lambda x_0 + z$ and it is clear that $z \in$ ker f.

Suppose $y = \lambda x_0 + z$ and $y = \lambda' x_0 + z'$. Then $(\lambda - \lambda') x_0 = z' - z$. If $\lambda' = \lambda$ then $z' = z$, but if $\lambda' \neq \lambda$ then $x_0 = \dfrac{z' - z}{\lambda - \lambda'}$ which contradicts the fact that $x_0 \notin$ ker f. $\square$

7.19 Theorem. *In a normed linear space $(X, \| \cdot \|)$, the kernel of a non-zero linear functional is either closed or dense.*

Proof. If ker f is closed then since f is non-zero, ker f is not dense. If ker f is not closed then there exists an x_0 a cluster point of ker f but where $x_0 \notin$ ker f. Now by Theorem 4.33, $\overline{\text{ker f}}$ is also a linear subspace of X and so $\overline{\text{ker f}}$ contains the span of x_0 and ker f. But by Lemma 7.18, X is the span of x_0 and ker f. So $X = \overline{\text{ker f}}$; that is, ker f is dense in X. $\square$

We can make the following deduction from Theorems 7.16 and 7.19.

7.20 Corollary. *Given a normed linear space $(X, \| \cdot \|)$, a linear functional f on X is continuous if and only if ker f is not dense in X.*

It is useful to consider a boundedness characterisation for the continuity of the inverse of a one-to-one linear mapping.

7.21 Theorem. *Given normed linear spaces $(X, \| \cdot \|)$ and $(Y, \| \cdot \|')$ and a linear mapping $T : X \to Y$, the inverse mapping T^{-1} exists and is a continuous linear mapping on $T(X)$ if and only if there exists an $m > 0$ such that*

$$m \|x\| \leq \|Tx\|' \quad for\ all\ x \in X.$$

<u>Proof.</u> Suppose that there exists an $m > 0$ such that

$$m\|x\| \leq \|Tx\|' \quad \text{for all } x \in X.$$

It follows that ker $T = \{\underline{0}\}$ and so T is one-to-one and T^{-1} exists on $T(X)$ and is clearly linear.

We show that T^{-1} is continuous on $T(X)$: Writing $x = T^{-1}y$ for any $y \in T(X)$, we have

$$m\|T^{-1}y\| \leq \|T(T^{-1}y)\|' = \|y\|'$$

so $\qquad \|T^{-1}y\| \leq \frac{1}{m}\|y\|' \quad \text{for all } y \in T(X)$

and therefore by Theorem 7.13, T^{-1} is continuous on $T(X)$.

Conversely, if T^{-1} is a continuous linear mapping on $T(X)$, then by Theorem 7.13 there exists an $M > 0$ such that

$$\|T^{-1}y\| \leq M\|y\|' \quad \text{for all } y \in T(X),$$

and since $y = Tx$,

$$\|T^{-1}(Tx)\| \leq M\|Tx\|';$$

that is, $\qquad \frac{1}{M}\|x\| \leq \|Tx\|' \quad \text{for all } x \in X. \ \square$

We are led naturally to examining linear mappings which are also homeomorphisms.

<u>7.22 Definition.</u> Given normed linear spaces $(X, \|\cdot\|)$ and $(Y, \|\cdot\|')$ a linear mapping $T : X \to Y$ is said to be a *topological isomorphism* (or a *linear homeomorphism*) if T is also a homeomorphism.

Normed linear spaces $(X, \|\cdot\|)$ and $(Y, \|\cdot\|')$ are said to be *topologically isomorphic* if there exists a topological isomorphism T of X onto Y.

<u>7.23 Remark.</u> Normed linear spaces which are topologically isomorphic under the topological isomorphism T are isomorphic as linear spaces under T and because T is a homeomorphism onto it follows from Theorems 6.27 and 6.29 that convergence of sequences is preserved under T and closed sets are preserved under T.

The relation "being topologically isomorphic to" is an equivalence relation on the set of all normed linear spaces. $\square$

From Theorem 7.13 and 7.21 we deduce the following convenient characterisation of a topological isomorphism.

7.24 Theorem. *Given normed linear spaces* $(X, \|\cdot\|)$ *and* $(Y, \|\cdot\|')$, *a linear mapping* $T : X \to Y$ *is a topological isomorphism if and only if there exist* m *and* $M > 0$ *such that*

$$m\|x\| \leq \|Tx\|' \leq M\|x\| \quad \text{for all } x \in X.$$

7.25 Example. Given a normed linear space $(X, \|\cdot\|)$ and a scalar $\alpha_0 \neq 0$, the *scale* mapping $T_{\alpha_0} : X \to X$ defined by

$$T_{\alpha_0}(x) = \alpha_0 x$$

is an obvious topological isomorphism of X onto X. $\square$

7.26 Remark. It is clear from Theorem 3.23 giving a characterisation for equivalent norms by inequalities, that if $\|\cdot\|$ and $\|\cdot\|'$ are equivalent norms for X then $(X, \|\cdot\|)$ and $(X, \|\cdot\|')$ are topologically isomorphic under the identity mapping. We have confirmation of Remark 4.17 from Remark 7.23 that closed sets are preserved under equivalent norms.

It also follows from Remark 6.28 that if T is a topological isomorphism of $(X, \|\cdot\|)$ into $(Y, \|\cdot\|')$ then $\|\cdot\|$ and $\|T.\|'$ are equivalent norms for X. So we can apply Theorem 3.40 to establish that a normed linear space which is topologically isomorphic to a Banach space is itself a Banach space. $\square$

We now examine further the nature of finite dimensional normed linear spaces. The next theorem explains why most analysis and linear algebra questions on finite dimensional normed linear spaces resolve themselves into a discussion of Euclidean (or Unitary) spaces.

7.27 Theorem. *All* m-*dimensional normed linear spaces over* $\mathbb{R}$ *(or* $\mathbb{C}$*) are topologically isomorphic to* $(\mathbb{R}^m, \|\cdot\|_2)$ *(or* $(\mathbb{C}^m, \|\cdot\|_2)$*).*

Proof. Consider an m-dimensional normed linear space $(X_m, \|\cdot\|)$ over $\mathbb{R}$ with basis $\{e_1, e_2, \ldots, e_m\}$. The mapping $T : X_m \to \mathbb{R}^m$ defined for $x \equiv \lambda_1 e_1 + \lambda_2 e_2 + \ldots + \lambda_m e_m$ by

$$Tx = (\lambda_1, \lambda_2, \ldots, \lambda_m)$$

is an isomorphism of X_m onto $\mathbb{R}^m$. But from Theorem 7.10, T is a topological isomorphism. $\square$

7.28 Remark. This isomorphism is used to translate any discussion of linear mappings on an m-dimensional linear space into matrix theory on $\mathbb{R}^m$ (or $\mathbb{C}^m$). Now we see how, for such a discussion we may choose the Euclidean (or Unitary) norm $\|\cdot\|_2$ as an appropriate norm on $\mathbb{R}^m$ (or $\mathbb{C}^m$) matching as it does our geometric intuition. $\square$

7.29 Remark. Noting Remark 7.26 and using Theorem 7.27 and Theorem 3.35 that Euclidean spaces are complete, we have independent confirmation of Theorem 3.36 that all finite dimensional normed linear spaces are complete. $\square$

From Remark 7.23 we make the following deduction.

7.30 Corollary. *All m-dimensional normed linear spaces over the same scalar field are topologically isomorphic.*

In Section 6 we saw that a fundamental class of mappings between metric spaces is the class of isometries. We noted in Remark 6.23 that if two metric spaces are isometric as metric spaces then they are identical as far as all the metric space properties are concerned. The analogous fundamental class of mappings for normed linear spaces is the class of isometric isomorphisms.

7.31 Definition. Given normed linear spaces $(X, \|\cdot\|)$ and $(Y, \|\cdot\|')$, a linear mapping $T : X \to Y$ is said to be an *isometric isomorphism* if

$$\|Tx\|' = \|x\| \quad \text{for all } x \in X.$$

Normed linear spaces $(X, \|\cdot\|)$ and $(Y, \|\cdot\|')$ are said to be *isometrically isomorphic* if there exists an isometric isomorphism T of X onto Y.

7.32 Remarks. Since T is linear it is clear that

$$\|Tx - Tx'\|' = \|T(x-x')\|' = \|x-x'\| \quad \text{for all } x,x' \in X$$

so an isometric isomorphism is a metric space isometry. But an isometric isomorphism is also a linear isomorphism which is also norm preserving. If two normed linear spaces are isometrically isomorphic then as normed linear spaces they are identical as far as all normed linear space properties are concerned.

The relation "being isometrically isomorphic to" is an equivalence relation on the set of all normed linear spaces. It is clear from Theorem 7.24 that an isometric isomorphism is also a topological isomorphism but obviously not every topological isomorphism is an isometric isomorphism. So the partition of normed linear spaces into topological isomorphism equivalence classes is coarser than the partition of normed linear spaces into isometrically isomorphic equivalence classes. □

7.33 Example. A common example of an isometric isomorphism is the natural identification of a normed linear space with the linear subspace of another normed linear space:

$(\mathbb{R}, |\cdot|)$ is isometrically isomorphic to the linear subspaces

(i) $\{(\lambda, 0) : \lambda \in \mathbb{R}\}$ of $(\mathbb{R}^2, \|\cdot\|_2)$ and

(ii) $\{(\lambda, \frac{\lambda}{2}) : \lambda \in \mathbb{R}\}$ of $(\mathbb{R}^2, \|\cdot\|_\infty)$.

In (i) the isometric isomorphism is $\lambda \mapsto (\lambda, 0)$ where $|\lambda| = \|(\lambda, 0)\|_2$ and in (ii) the isometric isomorphism is $\lambda \mapsto (\lambda, \frac{\lambda}{2})$ where $|\lambda| = \|(\lambda, \frac{\lambda}{2})\|_\infty$. Notice that the names of the elements are different but otherwise the spaces are identical as normed linear spaces. □

7.34 Example. Given a normed linear space $(X, \|\cdot\|)$ over $\mathbb{C}$ and a complex number λ_0 where $|\lambda_0| = 1$, the elementary *rotation* mapping $T_{\lambda_0} : X \to X$ defined by

$$T_{\lambda_0} x = \lambda_0 x$$

is linear and $\|T_{\lambda_0} x\| = \|\lambda_0 x\| = \|x\|$ for all $x \in X$. So T_{λ_0} is an isometric isomorphism on X.

In Example 6.24 we introduced the translation mapping as an isometry. Both translation and rotation mappings are isometries but translation is not a linear mapping so is not an isometric isomorphism. ☐

In Example 7.8 we saw that differentiation although a linear mapping is not continuous under the uniform norm. Nevertheless, differentiation does have a property closely related to continuity. The classical uniform convergence theorem for differentiation reads as follows.

7.35 Theorem. *If a sequence of continuously differentiable functions* $\{f_n\}$ *on* $[a,b]$ *is pointwise convergent to* f *on* $[a,b]$ *and is such that the sequence* $\{f_n'\}$ *is uniformly convergent on* $[a,b]$ *then*

(i) f is differentiable on $[a,b]$ and

(ii) $\displaystyle\lim_{n\to\infty} \frac{d}{dx} f_n(x) = \frac{d}{dx} \lim_{n\to\infty} f_n(x)$ for all $x \in [a,b]$.

Proof. Since $\{f_n'\}$ is a sequence in $C[a,b]$ uniformly convergent to some function g on $[a,b]$, we have since $(C[a,b], \|\cdot\|_\infty)$ is complete, (see Example 4.22), that $g \in C[a,b]$. But since integration is a continuous linear functional on $(C[a,b], \|\cdot\|_\infty)$, (see Example 7.5), we have for each $x \in [a,b]$ that

$$\int_a^x g(t)dt = \lim_{n\to\infty} \int_a^x f_n'(t)dt$$

$$= \lim_{n\to\infty} \left(f_n(x) - f_n(a) \right)$$

$$= f(x) - f(a).$$

But since $g \in C[a,b]$ we have from the Fundamental Theorem of Calculus that f is differentiable on $[a,b]$ and

$$g(x) = f'(x) \quad \text{for all } x \in [a,b];$$

that is,

$$\lim_{n \to \infty} \frac{d}{dx} f_n(x) = \frac{d}{dx} \lim_{n \to \infty} f_n(x) \quad \text{for all } x \in [a,b]. \; \square$$

Although in applying this theorem we need check no more than pointwise convergence of $\{f_n\}$, in fact the uniform convergence condition on $\{f_n'\}$ implies that the convergence of $\{f_n\}$ is uniform.

7.36 Corollary. *Under the conditions of Theorem 7.35, the sequence* $\{f_n\}$ *is uniformly convergent to f.*

Proof. In the proof of Theorem 7.35 we see that $\{f_n'\}$ is convergent to f' in $(C[a,b], \|\cdot\|_\infty)$. So given $\varepsilon > 0$ there exists a $\nu \in \mathbb{N}$ such that

$$\|f_n' - f'\|_\infty < \varepsilon \quad \text{for all } n > \nu.$$

Then for any $x \in [a,b]$,

$$\int_a^x |(f_n' - f')(t)| \, dt < \varepsilon(b-a) \quad \text{for all } n > \nu,$$

and so $|f_n(x) - f(x)| < \varepsilon(b-a) + |f_n(a) - f(a)| \quad \text{for all } n > \nu.$

Since $\{f_n\}$ is pointwise convergent to f there exists a $\nu_1 \in \mathbb{N}$ such that

$$|f_n(a) - f(a)| < \varepsilon \quad \text{for all } n > \nu_1.$$

Therefore,

$$|f_n(x) - f(x)| < \varepsilon(1+b-a) \quad \text{for all } n > \max(\nu, \nu_1);$$

that is, $\|f_n - f\|_\infty \leqslant \varepsilon(1+b-a) \quad \text{for all } n > \max(\nu, \nu_1). \; \square$

Corollary 7.36 shows that Theorem 7.35 can be interpreted to state that differentiation has a closed graph, (see Exercise 6.39.10(ii)).

7.37 Corollary. *The linear mapping* D *from* $(C^1[a,b], \|\cdot\|_\infty)$ *into* $(C[a,b], \|\cdot\|_\infty)$ *defined by*

$$D(f) = f'$$

has the property that for every sequence $\{f_n\}$ *converging to* f *in* $(C^1[a,b], \|\cdot\|_\infty)$ *where* $\{Df_n\}$ *converges to* g *in* $(C[a,b], \|\cdot\|_\infty)$ *we have* $g = Df$.

7.38 Remark. Notice that the linear functional D_0 on $(C^1[-\pi,\pi], \|\cdot\|_\infty)$ of Example 7.8 does not have a closed graph; (see Exercise 7.39.10(ii)). $\square$

7.39 Exercises.

1.   (i)   Prove that the functional I on $C[0,1]$ defined by

$$I(f) = \int_0^1 f(t)dt$$

is a continuous linear functional on $(C[0,1], \|\cdot\|_1)$.

(ii)   Determine whether the following linear mappings are continuous on $(C[0,1], \|\cdot\|_\infty)$ and $(C[0,1], \|\cdot\|_1)$.

(a)   For any given $t_0 \in [0,1]$, the linear functional $\hat{t}_0$ is defined by

$$\hat{t}_0(f) = f(t_0).$$

(b)   For any given $f_0 \in C[0,1]$, the linear functional I_{f_0} is defined by

$$I_{f_0}(f) = \int_0^1 f.f_0(t)dt.$$

(c)   For any given $f_0 \in C[0,1]$, the linear operator T_{f_0} is defined by

$$T_{f_0}(f) = f.f_0$$

2. (i)   Prove that the kth truncation mapping t_k from $(m, \|\cdot\|_\infty)$ into
            $(E_0, \|\cdot\|_\infty)$ defined by

$$t_k(\{\lambda_1, \lambda_2, \ldots, \lambda_n, \ldots\}) = \{\lambda_1, \lambda_2, \ldots, \lambda_k, 0, \ldots\}$$

            is a continuous linear mapping.

    (ii)   Prove that the shift operator S on m defined by

$$S(\{\lambda_1, \lambda_2, \ldots, \lambda_n, \ldots\}) = \{0, \lambda_1, \lambda_2, \ldots, \lambda_n, \ldots\}$$

            is an isometric isomorphism of m onto a subspace of m with
            respect to the supremum norm.

    (iii)   Consider the truncation operator t_k and the shift operator S
            on E_0. Prove that both are continuous as mappings from
            $(E_0, \|\cdot\|_2)$ into $(E_0, \|\cdot\|_\infty)$ and determine whether either is a
            topological isomorphism.

    (iv)   Prove that the normed linear space $(C[a,b], \|\cdot\|_\infty)$ is
            isometrically isomorphic to the normed linear space
            $(C[0,1], \|\cdot\|_\infty)$ under the mapping $f \longmapsto f^*$ where

$$f^*(t) = f\big(ta + (1-t)b\big) \quad \text{for all } t \in [0,1]$$

3.   Consider the linear space $\widetilde{C}(2\pi)$ of continuous real periodic
functions of period 2π on $\mathbb{R}$ and the linear space $C(\Gamma)$ where Γ is the unit
circle $\{(\lambda, \mu) \in \mathbb{R}^2 : \lambda^2 + \mu^2 = 1\}$.

    (i)   Prove that the normed linear space $(\widetilde{C}(2\pi), \|\cdot\|_\infty)$ is isometrically
            isomorphic to the normed linear space $(C(\Gamma), \|\cdot\|_\infty)$ under the
            mapping $f \longmapsto f^*$ where

$$f^*(\cos\theta, \sin\theta) = f(\theta) \quad \text{for all } \theta \in \mathbb{R}.$$

    (ii)   Deduce that

            (a) $(\widetilde{C}(2\pi), \|\cdot\|_\infty)$ is complete, and

            (b) $(\widetilde{C}(2\pi), \|\cdot\|_\infty)$ is a closed linear subspace of $(C(\mathbb{R}), \|\cdot\|_\infty)$
                where $C(\mathbb{R})$ is the linear space of bounded continuous real
                functions on $\mathbb{R}$.

4. (i)   Consider the functional f on ℓ_1 defined for
          $x \equiv \{\lambda_1,\lambda_2,\ldots,\lambda_n,\ldots\}$ by

$$f(x) = \sum \frac{n-1}{n} \cdot \lambda_n$$

Prove that f is linear and determine whether f is continuous
with respect to any of the norms $\|\cdot\|_\infty$, $\|\cdot\|_1$, $\|\cdot\|_2$ on ℓ_1.

   (ii) Let $\{\alpha_n\}$ be a bounded sequence of real numbers. Consider the
         operator T on ℓ_1 defined for $x \equiv \{\lambda_1,\lambda_2,\ldots,\lambda_n,\ldots\}$ by

$$Tx = \{\alpha_1\lambda_1,\alpha_2\lambda_2,\ldots,\alpha_n\lambda_n,\ldots\}.$$

   (a) Prove that T is a continuous linear operator on $(\ell_1,\|\cdot\|_1)$.

   (b) Prove that T is one-to-one if and only if $\alpha_n \neq 0$ for all
        $n \in \mathbb{N}$.

   (c) Prove that T is onto and is a topological isomorphism if
        and only if

$$\inf\{|\alpha_n| : n \in \mathbb{N}\} > 0.$$

5. (i)   By examining the mapping $T : \mathbb{R}^2 \to \mathbb{R}^2$ defined by

$$T(\lambda,\mu) = \left(\frac{\lambda+\mu}{2}, \frac{\lambda-\mu}{2}\right)$$

or otherwise, prove that $(\mathbb{R}^2,\|\cdot\|_\infty)$ and $(\mathbb{R}^2,\|\cdot\|_1)$ are
isometrically isomorphic.

   (ii) Show that $(\mathbb{R}^2,\|\cdot\|_\infty)$ and $(\mathbb{R}^2,\|\cdot\|_2)$ are not isometrically
         isomorphic.

   (iii) A normed linear space $(X,\|\cdot\|)$ is said to be *rotund* if for
         any $x \neq y$, $\|x\| = \|y\| = 1$ we have $\|x+y\| < 2$.

         Prove that a normed linear space which is not rotund cannot
         be isometrically isomorphic to a normed linear space which
         is rotund.

6.   (i)   Is every isometry between normed linear spaces an isometric isomorphism? Prove or give a counter-example.

    (ii)   Is every linear mapping which is also an isometry between normed linear spaces an isometric isomorphism? Prove or give a counter-example.

   (iii)   Is every isometry between complex normed linear spaces which takes $\underset{\sim}{0}$ to $\underset{\sim}{0}$ an isometric isomorphism? Prove or give a counter-example.

   (iv)   Is every isometry from $(\mathbb{R}^2, \|\cdot\|_2)$ onto $(\mathbb{R}^2, \|\cdot\|_2)$ which takes $\underset{\sim}{0}$ to $\underset{\sim}{0}$ an isometric isomorphism? Prove or give a counter-example.

7.   (i)   Given a continuous linear functional f on a normed linear space $(X, \|\cdot\|)$ and a scalar α, prove that the coset $\{x \in X : f(x) = \alpha\}$ is a closed linear subspace of X.

    (ii)   Consider $(c, \|\cdot\|_\infty)$ the linear space of convergent sequences with the supremum norm.

        (a)   Prove that the functional f on c defined for $x \equiv \{\lambda_1, \lambda_2, \ldots, \lambda_n, \ldots\}$ where $\lambda_n \to \lambda$ as $n \to \infty$ by

$$f(x) = \lambda$$

          is a continuous linear functional on $(c, \|\cdot\|_\infty)$.

        (b)   Prove that the set c_1 of sequences which converge to 1 is a closed coset of $(c, \|\cdot\|_\infty)$.

   (iii)   Determine whether the set

$$\{f \in C[0,1] : \int_0^1 f(t)dt = 1\}$$

      is a closed coset of

      (a) $(C[0,1], \|\cdot\|_\infty)$ and

      (b) $(C[0,1], \|\cdot\|_1)$.

8. (i)   Let k be a continuous real function on the square
$$\Box \equiv \{(x,t) : a \leqslant x \leqslant b, \ a \leqslant t \leqslant b\}.$$
The *Fredholm operator* K on $C[a,b]$ is defined by

$$(Kf)(x) = \int_a^b k(x,t)f(t)dt.$$

Prove that K is a continuous linear operator on
$(C[a,b], \|\cdot\|_\infty)$;   (see 5.10.1).

(ii)   Let k be a continuous real function on the triangle
$$\Delta \equiv \{(x,t) : a \leqslant t \leqslant x, \ a \leqslant x \leqslant b\}.$$
The *Volterra operator* K on $C[a,b]$ is defined by

$$(Kf)(x) = \int_a^x k(x,t)f(t)dt.$$

Prove that K is a continuous linear operator on $(C[a,b], \|\cdot\|_\infty)$;
(see 5.10.3).

9.   Consider normed linear spaces $(X, \|\cdot\|)$ and $(Y, \|\cdot\|')$ and a
linear mapping $T : X \to Y$.

(i)   Prove that if the set $\{\|Tx_n\|' : n \in \mathbb{N}\}$ is bounded for every
sequence $\{x_n\}$ which is convergent to $\underset{\sim}{0}$ in $(X, \|\cdot\|)$ then T is
continuous.

(ii)   Suppose the mapping T has the property that $\{Tx_n\}$ is a Cauchy
sequence in $(Y, \|\cdot\|')$ if and only if $\{x_n\}$ is a Cauchy sequence
in $(X, \|\cdot\|)$. Prove then that T is a topological isomorphism.

10. (i)   Given normed linear spaces $(X, \|\cdot\|)$ and $(Y, \|\cdot\|')$ and a mapping
$T : X \to Y$, prove that if T has a closed graph then ker T is
closed;   (see Exercise 6.39.10(ii)).

(ii)   Deduce that a linear functional on a normed linear space is
continuous if it has a closed graph.

(iii)   In Remark 7.17 we saw that the identity mapping id from
$(C[0,1], \|\cdot\|_1)$ into $(C[0,1], \|\cdot\|_\infty)$ has a closed kernel but is not
continuous. Show that this same mapping has a closed graph and
so establish that a linear mapping between normed linear spaces
with a closed graph is not necessarily continuous.

10. (iv)   By examining the mapping T from $(\ell_1, \|\cdot\|_\infty)$ into $(c_0, \|\cdot\|_\infty)$
            defined by

$$T(\{\lambda_1, \lambda_2, \ldots, \lambda_n, \ldots\})$$
$$= \{\lambda_1 + \lambda_2 + \lambda_3 + \ldots, \frac{-\lambda_1 + \lambda_2 + \lambda_3 + \ldots}{2}, \frac{\lambda_1 - \lambda_2 + \lambda_3 + \lambda_4 + \ldots}{3}, \ldots\}$$

or otherwise, show that a linear mapping with a closed kernel
does not necessarily have a closed graph.

11. (i)   For the linear space $C^1[a,b]$ of continuously differentiable
            real functions on $[a,b]$ consider the norm

$$\|f\|' = \|f\|_\infty + \|f'\|_\infty .$$

(a)   Show that the norm $\|\cdot\|'$ is not equivalent to norm $\|\cdot\|_\infty$
       on $C^1[a,b]$.

(b)   Prove that the normed linear space $(C^1[a,b], \|\cdot\|')$ is
       complete.

(c)   Prove that the mapping D defined by

$$D(f) = f'$$

is continuous on $(C^1[a,b], \|\cdot\|')$.

(ii)   Consider the linear subspace $P_3[a,b]$ of cubic polynomials on
        $[a,b]$.

(a)   Prove that D is continuous on $(P_3[a,b], \|\cdot\|_\infty)$.

(b)   Prove that $\|\cdot\|_\infty$ and $\|\cdot\|'$ are equivalent norms on $P_3[a,b]$.

(iii)   Consider the linear subspace $P[a,b]$ of polynomials on $[a,b]$.

(a)   Show that D is not continuous on $(P[a,b], \|\cdot\|_\infty)$.

(b)   Show that $(P[a,b], \|\cdot\|')$ is not complete.

(c)   Show that $\|\cdot\|_\infty$ and $\|\cdot\|'$ are not equivalent norms on
       $P[a,b]$.

12. (i)   Prove that if a normed linear space $(X, \| \cdot \|)$ is topologically
          isomorphic to a separable normed linear space $(Y, \| \cdot \|')$ then
          $(X, \| \cdot \|)$ is also separable.

   (ii)   Deduce that the normed linear spaces $(c_0, \| \cdot \|_\infty)$ and $(m, \| \cdot \|_\infty)$
          are not topologically isomorphic.

13. (i)   A functional f on a linear space X is said to be *additive* if

          $$f(x+y) = f(x) + f(y) \quad \text{for all } x, y \in X.$$

          Prove that a continuous additive functional f on a real
          normed linear space is a linear functional.

   (ii)   A linear space X over $\mathbb{C}$ can be regarded as a linear space X
          over $\mathbb{R}$.

          (a)   Given a complex linear functional f on X, prove that Re f
                is a real linear functional on X and given a real linear
                functional $f_\mathbb{R}$ on X prove that a complex linear functional
                f on X is defined by

                $$f(x) = f_\mathbb{R}(x) - if_\mathbb{R}(ix).$$

          (b)   Prove that a complex linear functional f on $(X, \| \cdot \|)$ over
                $\mathbb{C}$ is continuous if and only if the real linear functional
                $f_\mathbb{R}$ on $(X, \| \cdot \|)$ over $\mathbb{R}$ is continuous.

   (iii) Prove that a continuous additive functional f on a complex
          normed linear space is a linear functional.

14.   Consider a normed linear space $(X, \| \cdot \|)$ over $\mathbb{C}$.

   (i)   Given $(X \times X, \| \cdot \|_\pi)$, (see Exercise 1.34.11) prove that the
          mapping $T : X \times X \to X$ defined by

          $$T(x,y) = x + y$$

          is continuous on $X \times X$; (see Exercise 6.39.10(i));
          (that is, addition of vectors is a jointly continuous operation).

14. (ii)   Given $(\mathbb{C} \times X, \|\cdot\|_\pi)$, prove that the mapping $T : \mathbb{C} \times X \to X$ defined by

$$T(\lambda, x) = \lambda x$$

is continuous on $\mathbb{C} \times X$;

(that is, multiplication by a scalar is a jointly continuous operation).

15.   Given normed linear spaces $(X, \|\cdot\|)$ and $(Y, \|\cdot\|')$ and a continuous linear mapping $T : X \to Y$, consider the quotient normed linear space $(\frac{X}{\ker T}, \|\cdot\|)$, (see Exercise 4.51.21).

  (i)   Prove that the quotient mapping $\pi : X \to \frac{X}{\ker T}$ defined by

$$\pi(x) = x + \ker T$$

is a continuous linear mapping.

  (ii)   Prove that there is a unique mapping $S : \frac{X}{\ker T} \to Y$ such that $T = S \circ \pi$, and that S is linear.

  (iii) Prove that S is continuous.

16.   Consider an incomplete normed linear space $(X, \|\cdot\|)$ and its completion $(\widetilde{X}, \|\cdot\|) \equiv (\frac{X^+}{\ker p^+}, \|\cdot\|^+)$, (see Exercise 3.53.15).

  (i)   Prove that the mapping $T : X \to \frac{X^+}{\ker p^+}$ defined by

$$Tx = [x]$$

where $[x]$ is the equivalence class of sequences $\{x_n\}$ in X which converge to x, is an isometric isomorphism of X into $\frac{X^+}{\ker p^+}$.

  (ii)   Prove that $T(X)$ is dense in $(\widetilde{X}, \|\cdot\|)$. (So $(X, \|\cdot\|)$ is identified with a dense linear subspace of its completion $(\widetilde{X}, \|\cdot\|)$ in a natural way.)

IV. COMPACTNESS

In the study of real analysis we easily isolate the property
known as the local compactness of $\mathbb{R}$ which we recall states that every
bounded sequence has a convergent subsequence. The local compactness of
$\mathbb{R}$ implies that bounded closed intervals are of special significance in
that every sequence $\{x_n\}$ in a bounded closed interval $[a,b]$ has a
subsequence $\{x_{n_k}\}$ which is convergent to a point in $[a,b]$.

A key application of this property in real analysis concerns
continuous functions defined on bounded closed intervals. Such functions
have the important properties of being bounded, attaining a maximum and
minimum on their domain and of being uniformly continuous.

We extend the compactness property to metric spaces by defining
compact subsets which generalise the compactness property of bounded
closed intervals in real analysis. We then explore the special properties
of continuous functions defined on such compact sets generalising the real
analysis applications.

It is the special properties of continuous functions on
bounded closed intervals which establishes the linear space $C[a,b]$ as a
subspace of $B[a,b]$, (see Example 1.15). We are now able to generalise
the normed linear space $(C[a,b], \|\cdot\|_\infty)$ to $(C(X), \|\cdot\|_\infty)$ the linear space of
continuous functions on the compact metric space (X,d) with the supremum
norm. The compactness property brings considerable insight into the
structural properties of such a normed linear space $(C(X), \|\cdot\|_\infty)$ through
the celebrated Weierstrass and Stone-Weierstrass Approximation Theorems
and the Ascoli-Arzelà Theorem.

8. SEQUENTIAL COMPACTNESS IN METRIC SPACES

Our analysis of metric spaces has been developed using con-
vergence of sequences as the primary tool. The local compactness of $\mathbb{R}$ is
a property expressed in terms of convergence of sequences. Our formula-
tion of compactness for metric spaces keeps the sequential approach and
is an obvious generalisation of the real situation.

<u>8.1 Definition</u>. Given a metric space (X,d), a subset A is said to be
compact if every sequence $\{x_n\}$ in A has a subsequence $\{x_{n_k}\}$ which is
convergent to a point of A.

<u>8.2 Remark</u>. As we shall see, there are other formulations of compactness
which are equivalent for metric spaces. For clarity, especially when
other forms are being discussed, compactness as in Definition 8.1 is
sometimes referred to as *sequential compactness*. ☐

The following elementary examples of compact sets should not
be overlooked.

<u>8.3 Example</u>. Given any metric space (X,d), any finite subset
$E \equiv \{x_k : k \in \{1,2,\ldots,n\}\}$ is a compact set: Any infinite sequence formed
from members of E has to have a subsequence which repeats on one term
and such a subsequence is clearly convergent to a point in E. ☐

<u>8.4 Example</u>. Given any metric space (X,d), any set E consisting of the
terms of a convergent sequence $\{x_n\}$ and its limit point x_0,
$E \equiv \{x_n : n \in \mathbb{N}\} \cup \{x_0\}$, is a compact set: Any sequence formed from
members of E either has a finite range in which case it is as in
Example 8.3 or it has an infinite range in which case it has as a
subsequence a subsequence of $\{x_n\}$ which of course converges to x_0 in E. ☐

To prove that a set is compact we need to prove that every
sequence in the set has a convergent subsequence. It is often a much
simpler task to show that a set is not compact, because we need only
exhibit one sequence which does not have a convergent subsequence.

<u>8.5 Examples</u>. In $\mathbb{R}$ with the usual metric, subsets (i) $(0,1]$ and (ii) $\mathbb{Z}$ are not compact:

   (i)   The sequence $\{\frac{1}{n}\}$ is convergent to $0 \notin (0,1]$, so all
           subsequences of $\{\frac{1}{n}\}$ are convergent to 0 and no subsequence
           is then convergent in $(0,1]$.

   (ii)   The sequence $\{n\}$ is unbounded and all subsequences of $\{n\}$ are
           unbounded so no subsequence is convergent. $\square$

<u>8.6 Example</u>. In the normed linear space $(c_0, \|\cdot\|_\infty)$, the unit sphere $S(0;1)$ is not compact: The sequence $\{x_n\}$ where

$$x_1 \equiv \{1,0,\ldots \qquad \}$$
$$x_2 \equiv \{0,1,0,\ldots \qquad \}$$
$$\cdot\ \cdot\ \cdot$$
$$x_n \equiv \{0,\ldots,0,1,0,\ldots\}$$
$$\text{nth place}$$
$$\cdot\ \cdot\ \cdot$$

has the property that $\|x_n - x_m\|_\infty = 1$ for all $m \neq n$, so $\{x_n\}$ does not have any convergent subsequence. $\square$

The cases in Examples 8.5 suggest that compact sets in general should have the following properties.

<u>8.7 Theorem</u>. *A compact subset of a metric space* (X,d) *is*
(i) bounded and (ii) closed.

<u>Proof</u>.   (i) Consider an unbounded subset A. Then given $x_0 \in A$, for each $n \in \mathbb{N}$ there exists an $x_n \in A$ such that $d(x_n, x_0) > n$. That is, the sequence $\{x_n\}$ in A is unbounded and every subsequence of $\{x_n\}$ is unbounded so none can be convergent. Therefore A is not compact.

        (ii) Consider a subset A which is not closed. Then there exists a cluster point x_0 of A in $X \setminus A$. This implies that there exists a sequence $\{x_n\}$ in A which is convergent to x_0. All subsequences of $\{x_n\}$ converge to x_0 and so none is convergent in A. Therefore A is not compact. $\square$

8.8 Remark. From this theorem it is clear that in $\mathbb{R}$ with the usual metric, the set $(0,1]$ of Example 8.5(i) is not compact because it is not closed and the set $\mathbb{Z}$ of Example 8.5(ii) is not compact because it is not bounded. However, we note that in general a bounded closed subset of a metric space is not necessarily compact. In Example 8.6 we saw that the unit sphere $S(0;1)$ of the normed linear space $(c_0, \|\cdot\|_\infty)$ is not compact although it is both bounded and closed. $\square$

Nevertheless, in finite dimensional normed linear spaces, compact sets do have a very simple characterisation.

8.9 Theorem. *In a finite dimensional normed linear space* $(X_m, \|\cdot\|)$, *a subset is compact if and only if it is closed and bounded.*

Proof. We need only prove that a bounded closed subset A is compact. Let $\{e_1, e_2, \ldots, e_m\}$ be a basis for X_m. Consider any sequence $\{x_n\}$ in A where

$$x_1 \equiv \lambda_1^1 e_1 + \lambda_2^1 e_2 + \ldots + \lambda_m^1 e_m$$

$$x_2 \equiv \lambda_1^2 e_1 + \lambda_2^2 e_2 + \ldots + \lambda_m^2 e_m$$

$$\cdot \ \cdot \ \cdot$$

$$x_n \equiv \lambda_1^n e_1 + \lambda_2^n e_2 + \ldots + \lambda_m^n e_m$$

$$\cdot \ \cdot \ \cdot$$

Since A is bounded the sequence $\{x_n\}$ is bounded. For $x \equiv \lambda_1 e_1 + \lambda_2 e_2 + \ldots + \lambda_m e_m$ the co-ordinate functionals f_k where $k \in \{1, 2, \ldots, m\}$, defined by

$$f_k(x) = \lambda_k$$

are continuous linear functionals on $(X_m, \|\cdot\|)$, (see Theorem 7.10), so each co-ordinate sequence $\{\lambda_n^k\}$ for $k \in \{1, 2, \ldots, m\}$ is a bounded sequence of scalars, (see Remark 7.14). By the local compactness of $\mathbb{R}$, $\{\lambda_n^1\}$ has a convergent subsequence. Consider the corresponding subsequence of $\{\lambda_n^2\}$. By the local compactness of $\mathbb{R}$, this has a convergent subsequence. Continuing this process to the nth co-ordinate sequence we obtain a subsequence $\{x_{n_k}\}$ which has each co-ordinate sequence convergent. Then by

Theorem 3.8, $\{x_{n_k}\}$ is convergent to some $x_0 \in X_m$. Then x_0 either belongs to A or is a cluster point of A. But since A is closed, $x_0 \in A$ and we conclude that A is compact. $\square$

      A normed linear space consists of an algebraic structure and a metric structure. Results which show that a property in one of these fundamental structures has a consequence in the other, are particularly fascinating. In Theorem 7.10 we saw that on a finite dimensional normed linear space the linearity of a mapping (an algebraic property) implies its continuity (a metric space property). But our next theorem is remarkable in that it is a case where a metric property has a unexpected algebraic implication.

      We now proceed to show that those normed linear spaces with the property that all bounded closed subsets are compact must necessarily be finite dimensional. To prove this we need a small technical lemma.

8.10 Riesz Lemma. *Let M be a proper closed linear subspace of a normed linear space* $(X, \|\cdot\|)$. *Then for each* $0 < \delta < 1$ *there exists an* $x_\delta \in X$ *where* $\|x_\delta\| = 1$ *such that* $d(x_\delta, M) \geq \delta$.

Proof. Since M is proper there exists an $x_1 \in C(M)$ and since M is closed, $d \equiv d(x_1, M) > 0$, (see Exercise 4.51.12). So there exists some $x_0 \in M$ such that $\|x_1 - x_0\| \leq \dfrac{d}{\delta}$.

Write $\quad x_\delta \equiv \dfrac{x_1 - x_0}{\|x_1 - x_0\|}$.

For all $x \in M$,

$$\|x_\delta - x\| = \frac{1}{\|x_1 - x_0\|} \left\| x_1 - (x_0 + \|x_1 - x_0\| x) \right\|$$

and since $x_0 + \|x_1 - x_0\| x \in M$ we have

$$\|x_\delta - x\| \geq d / \frac{\delta}{d} = \delta.$$

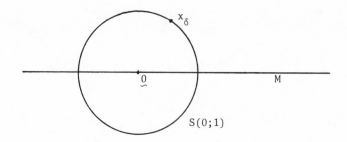

Figure 15. For $0 < \delta < 1$, $\|x_\delta\| = 1$ and $d(x_\delta, M) \geq \delta$. ☐

8.11 Remark. The Riesz Lemma states that for any proper closed linear
subspace M of a normed linear space there exist points in the unit
sphere $S(\underline{0};1)$ whose distance from M is arbitrarily close to 1. However,
it is not true in general that there is a point in the sphere $S(\underline{0};1)$ whose
distance from M is actually 1; (see Exercise 8.41.17(ii)). ☐

8.12 Riesz Theorem. *A normed linear space is finite dimensional if and*
only if the closed unit ball is compact.

Proof. From Theorem 8.9 it is clear that in a finite dimensional normed
linear space the closed unit ball is compact.

Conversely, for any infinite dimensional normed linear space
$(X, \|\cdot\|)$ we show that the closed unit ball is not compact:
For any $x_1 \in X$ where $\|x_1\| = 1$ consider $M_1 \equiv sp\{x_1\}$. Since M_1 is finite
dimensional it is a proper closed linear subspace of $(X, \|\cdot\|)$, (see
Corollary 4.21). By Riesz Lemma there exists an $x_2 \in C(M_1)$ where
$\|x_2\| = 1$ such that $d(x_2, M_1) \geq \frac{1}{2}$. Consider $M_2 \equiv sp\{x_2, M_1\}$. Again M_2 is
finite dimensional and so is a proper closed linear subspace of $(X, \|\cdot\|)$.
Since $(X, \|\cdot\|)$ is infinite dimensional we can continue this process to
obtain a sequence $\{x_n\}$ where $\|x_n\| = 1$ for all $n \in \mathbb{N}$ and

$$\|x_n - x_m\| \geq \frac{1}{2} \quad \text{for all } m \neq n.$$

Then $\{x_n\}$ is a sequence which cannot have any convergent subsequence. But
$\{x_n\}$ is contained in the closed unit ball $B[\underline{0};1]$ which is therefore not
compact. ☐

Although in general bounded closed subsets are not necessarily compact we show in the next theorem that closed subsets of compact sets are compact.

8.13 Theorem. *Given a metric space* (X,d) *and a compact subset* A, *a closed subset* F *of* A *is compact.*

Proof. Consider a sequence $\{x_n\}$ in F. Now $\{x_n\}$ is a sequence in A. But A is compact so $\{x_n\}$ has a subsequence $\{x_{n_k}\}$ which is convergent to some $x \in A$. Then $x \in F$ or x is a cluster point of F. But F is closed so $x \in F$. Therefore, $\{x_{n_k}\}$ is convergent to $x \in F$; that is, F is compact. $\square$

This theorem provides a useful method for deducing the compactness or otherwise of some subsets.

The significance of the compactness concept is illustrated by the property that compactness is invariant under continuous mappings.

8.14 Theorem. *Given metric spaces* (X,d) *and* (Y,d') *and a continuous function* $T : X \to Y$, *if* A *is a compact subset of* X *then* T(A) *is a compact subset of* Y.

Proof. Consider a sequence $\{y_n\}$ in T(A) and a sequence $\{x_n\}$ in A where $y_n = Tx_n$ for each $n \in \mathbb{N}$. Now A is compact so $\{x_n\}$ has a subsequence $\{x_{n_k}\}$ convergent to some $x \in A$. But T is continuous at x so $\{y_{n_k}\}$, where $y_{n_k} = Tx_{n_k}$ for each $k \in \mathbb{N}$, is convergent to Tx. But $\{y_{n_k}\}$ is a subsequence of $\{y_n\}$ and is convergent to a point of T(A); that is, T(A) is compact. $\square$

8.15 Remark. From Theorems 8.13 and 8.14 we can see more clearly how the compactness or otherwise of the closed unit ball determines the characterisation of compactness in normed linear spaces:

Any bounded closed set in a normed linear space $(X, \|\cdot\|)$ is a closed subset of the closed ball $B[\underline{0};r]$ for some $r > 0$. But $B[\underline{0};r]$ is the image of $B[\underline{0};1]$ under the continuous scale mapping T_r on X, (see Example 7.25). So if $B[\underline{0};1]$ is compact then every bounded closed subset is compact.

Now any closed ball $B[x;r]$ in a normed linear space $(X, \|\cdot\|)$ is the image under a homeomorphism which is the composition of a translation mapping, (see Example 6.24), and a scale mapping. So we notice that in an infinite dimensional normed linear space, because the closed unit ball is not compact, a compact subset cannot contain any closed ball $B[x;r]$. $\square$

Theorem 8.14 has a particularly useful special case.

8.16 Corollary. *Given a metric space* (X,d), *a non-empty compact subset* A *and a continuous real function* f *defined on* A *then* $f(A)$ *is bounded and has a maximum and minimum on* A.

Proof. From Theorem 8.14, $f(A)$ is compact so is bounded and closed. Since $f(A)$ is bounded as a subset of $\mathbb{R}$ with the usual norm, sup $f(A)$ and inf $f(A)$ exist and since $f(A)$ is closed both belong to $f(A)$ and are therefore max $f(A)$ and min $f(A)$. $\square$

This corollary is applied in what is called approximation theory. In a metric space, given an element we want to find a "best approximation" to that element from a subset defined by a particular property. By "best approximation" we mean that we want to find the closest element in the subset to our given element; "closest" relates to the metric we impose on the space. In the case where the approximating element is to be chosen from a compact subset, we have the following general existence result.

8.17 Corollary. *Given a metric space* (X,d) *and a non-empty compact subset* A, *to every* $x_0 \notin$ A *there exists a* $y_0 \in$ A *such that*

$$d(x_0,y_0) = d(x_0,A).$$

Proof. Consider the real function f on A defined by

$$f(x) = d(x,x_0), \quad \text{(see Exercise 6.39.9(i)(a)).}$$

Now $|f(x)-f(y)| \leqslant d(x,y)$, (see Exercise 1.34.8), so f is continuous on A. But A is compact so f has a minimum on A; that is, there exists a $y_0 \in$ A such that

$$f(y_0) = d(x_0,y_0) = \inf\{d(x_0,y) : y \in A\}$$
$$= d(x_0,A). \ \square$$

<u>8.18 Remark</u>. Notice that when A is not compact a "best approximating" element need not exist: In $\mathbb{R}$ with the usual norm, for $A \equiv (0,1)$ and $x_0 = 2$ we have $d(2,A) = 1$ but there is no closest point in A to 2. $\square$

A significant class of such best approximation problems is addressed by the following general result.

<u>8.19 Theorem</u>. *Given a normed linear space* $(X, \|\cdot\|)$ *and a finite dimensional linear subspace* X_m, *for any* $x_0 \notin X_m$ *there exists a* $y_0 \in X_m$ *such that*

$$\|x_0 - y_0\| = d(x_0, X_m).$$

<u>Proof</u>. Writing $d \equiv d(x_0, X_m)$ consider the closed ball $B[x_0; 2d]$. Now $B[x_0; 2d] \cap X_m$ is a closed and bounded non-empty subset of $(X_m, \|\cdot\|\big|_{X_m})$, (see Exercise 4.51.8(i)). Since X_m is finite dimensional we deduce from Riesz Theorem 8.12 that $B[x_0; 2d] \cap X_m$ is compact. Then from Corollary 8.17 there exists a $y_0 \in B[x_0; 2d] \cap X_m$ such that

$$\|x_0 - y_0\| = d(x_0, B[x_0; 2d] \cap X_m)$$
$$= d(x_0, X_m).$$

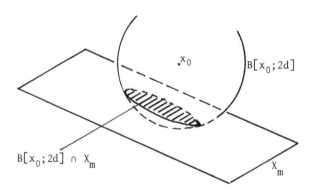

Figure 16.   The ball $B[x_0; 2d]$ "scoops" a compact
set out of X_m. $\square$

<u>8.20 Example</u>. For each continuous real function on $[a,b]$ there exist a cubic polynomials on $[a,b]$ which are "best fit" in the sense that they
(i) minimise the ordinate difference or (ii) minimise the area difference:

Now we are considering the linear space $C[a,b]$ and the four dimensional linear subspace $P_3[a,b]$ of cubic polynomials on $[a,b]$. In (i) we are considering best approximation under the supremum norm $\|\cdot\|_\infty$ and in (ii) under the integral norm $\|\cdot\|_1$. Theorem 8.19 guarantees the existence of such best approximations but an approximating cubic is not necessarily the same in both cases nor is it unique in either case. $\square$

In a normed linear space, rotundity is a condition sufficient to guarantee uniqueness of approximation in convex sets, (see Exercise 7.39.5(iii)).

8.21 Definition. A normed linear space $(X, \|\cdot\|)$ is said to be *rotund* if for any $x \neq y$, $\|x\| = \|y\| = 1$ we have $\|x+y\| < 2$.

8.22 Theorem. *Given a rotund normed linear space $(X, \|\cdot\|)$ and a convex subset A, if for $x_0 \notin A$ there exists a $y_0 \in A$ such that $\|x_0-y_0\| = d(x_0,A)$ then y_0 is unique.*

Proof. Suppose there exist y_0 and y_0' in A such that
$$d(x_0,A) = \|x_0-y_0\| = \|x_0-y_0'\|.$$
Since A is convex, $\dfrac{y_0+y_0'}{2} \in A$. Then

$$d(x_0,A) = \tfrac{1}{2}(\|x_0-y_0\| + \|x_0-y_0'\|)$$

$$\geq \left\|x_0 - \frac{y_0+y_0'}{2}\right\| \geq d(x_0,A).$$

So $\|x_0-y_0\| + \|x_0-y_0'\| = \left\|x_0 - \dfrac{y_0+y_0'}{2}\right\|$

Since $(X, \|\cdot\|)$ is rotund we deduce that $y_0' = y_0$. $\square$

8.23 Remark. Our study of compactness implying the existence of best approximating elements does not provide a method of actually calculating those elements in particular cases. Nevertheless, it does establish the first basic step in our approach to approximation theory. $\square$

In our study of compactness we have so far worked exclusively with the sequential form of compactness and we have seen that this is sufficient to cope with many applications. Yet the sequential form of

compactness does not reveal the underlying theoretical advantages of the
compactness property in analysis: Compactness is a way of reducing an
infinite process to a finite process. To reveal this aspect of compactness
we examine alternative forms of compactness.

8.24 Definitions. For a subset A of a metric space (X,d), given $\varepsilon > 0$, a
finite subset $E \equiv \{x_1, x_2, \ldots, x_n\}$ of A is called an ε-*net* for A if
$A \subseteq \cup \{B(x_k; \varepsilon) : k \in \{1, 2, \ldots, n\}\}$.
The subset A is said to be *totally bounded* if for any given $\varepsilon > 0$, there
exists an ε-net for A; that is, if for any $\varepsilon > 0$, A can be covered by a
finite union of open balls of radius ε with centres in A.

8.25 Remark. It is clear that for any subset A of a metric space (X,d),
if there exists an $\varepsilon > 0$ and an ε-net for A then A is bounded. So a
totally bounded set is bounded. However the converse is not true in
general: In Example 8.6 we considered the unit sphere $S(0;1)$ in the
normed linear space $(c_0, \|\cdot\|_\infty)$ and the sequence $\{x_n\}$ where

$$x_1 \equiv \{1, 0, \ldots \qquad \}$$
$$x_2 \equiv \{0, 1, 0, \ldots \qquad \}$$
$$\cdot \ \cdot \ \cdot$$
$$x_n \equiv \{0, \ldots, 0, 1, 0, \ldots\}$$
$$\text{nth place}$$
$$\cdot \ \cdot \ \cdot$$

Now $\|x_n - x_m\|_\infty = 1$ for all $m \neq n$.
So for $\varepsilon \leqslant \tfrac{1}{2}$, any open ball with radius ε which contains a particular
element of the sequence $\{x_n\}$ contains no other element of the sequence.
So $S(0;1)$ does not have an ε-net for $\varepsilon \leqslant \tfrac{1}{2}$ and is therefore not totally
bounded. $\square$

The example in Remark 8.25 suggests a link between sequential
compactness and total boundedness. In exploring this link we make use of
the fact that both sequential compactness and total boundedness are
intrinsic properties of a set like completeness and unlike closedness which
is relative to the space in which the set lies. This follows immediately
from the Definitions 8.1 and 8.24.

8.26 Remark. Given a subset A of a metric space (X,d),

   (i)   A is sequentially compact if and only if the metric subspace
         $(A,d|_A)$ is sequentially compact, and

   (ii) A is totally bounded if and only if the metric subspace
         $(A,d|_A)$ is totally bounded. $\square$

In considering the relations between the different forms of compactness it is sufficient to work from a sequentially compact metric space and in doing so we avoid many small distractions from the main argument.

8.27 Theorem. *A sequentially compact metric space* (X,d) *is complete and totally bounded.*

Proof. Since the metric space (X,d) is sequentially compact every Cauchy sequence $\{x_n\}$ in X has a convergent subsequence and so $\{x_n\}$ is convergent, (see Exercise 3.53.5); that is, (X,d) is complete.
Suppose that (X,d) is not totally bounded. Then for some $\varepsilon_0 > 0$, there is no ε_0-net for X. We define a sequence $\{x_n\}$ in X as follows: Choose any $x_1 \in X$. Since there is no ε_0-net, $C\big(B(x_1;\varepsilon_0)\big) \neq \phi$ so choose $x_2 \in C\big(B(x_1;\varepsilon_0)\big)$. Again $C\big(B(x_1;\varepsilon_0) \cup B(x_2;\varepsilon_0)\big) \neq \phi$ so choose $x_3 \in C\big(B(x_1;\varepsilon_0) \cup B(x_2;\varepsilon_0)\big)$. Continuing inductively we have a sequence $\{x_n\}$ where $d(x_n,x_m) \geq \varepsilon_0$ for all $m \neq n$. So $\{x_n\}$ cannot have a convergent subsequence and therefore (X,d) is not sequentially compact. $\square$

Theorem 8.27 adds to the information of Theorem 8.7 as follows.

8.28 Corollary. *In a metric space a compact subset is closed and totally bounded.*

We can now deduce that the boundedness and total boundedness concepts coincide in finite dimensional normed linear spaces.

8.29 Corollary. *In a finite dimensional normed linear space* $(X_m, \|\cdot\|)$, *a subset is totally bounded if and only if it is bounded.*

Proof. In view of Remark 8.25 we need only prove that boundedness implies total boundedness and in view of Remark 8.15 it is sufficient to confine our attention to the closed unit ball. Theorem 8.9 gives us that the closed unit ball is sequentially compact so by Theorem 8.27 it is totally bounded. □

But further the equivalence of boundedness and total boundedness characterises finite dimensionality in normed linear spaces and we have an extension of Riesz Theorem 8.12.

8.30 Corollary. *A normed linear space* $(X, \|\cdot\|)$ *is finite dimensional if and only if the closed unit ball is totally bounded.*

Proof. If $(X, \|\cdot\|)$ is finite dimensional the total boundedness of the closed unit ball is given by Corollary 8.29.

If $(X, \|\cdot\|)$ is infinite dimensional, then examining the sequence $\{x_n\}$ constructed in the proof of Riesz Theorem 8.12 we notice that $\|x_n - x_m\| \geq \frac{1}{2}$ for all $m \neq n$. So for $\varepsilon \leq \frac{1}{4}$, any open ball with radius ε which contains a particular element of the sequence $\{x_n\}$ contains no other element of the sequence. So $B[\underline{0};1]$ does not have an ε-net for $\varepsilon \leq \frac{1}{4}$ and therefore is not totally bounded. □

It is often convenient to express compactness in what is apparently a more general form.

8.31 Definitions. Consider a metric space (X,d) and a subset A. A family of subsets $\{E_\alpha\}$ is called a *cover* for A if $A \subseteq \cup E_\alpha$. Any subfamily of $\{E_\alpha\}$ which is itself a cover for A is called a *subcover* for A.
The subset A is said to be *ball cover compact* if every cover for A by open balls with centres in A has a finite subcover.

8.32 Example. It is easy to see that the unit sphere $S(\underline{0};1)$ in the normed linear space $(c_0, \|\cdot\|_\infty)$ discussed in Example 8.6 and Remark 8.25 is not ball cover compact: From an argument similar to that used in Remark 8.25 we deduce that the cover by open balls $\{B(x;\frac{1}{2}) : x \in S(\underline{0};1)\}$ has no finite subcover. □

The intrinsic character of ball cover compactness is an immediate implication of Definition 8.31.

8.33 Remark. Given a subset A of a metric space (X,d), A is ball cover compact if and only if the metric subspace $(A,d|_A)$ is ball cover compact. $\square$

We note that ball cover compactness is explicit in reducing a possibly infinite cover to a finite subcover. We might expect that ball cover compactness is related to total boundedness. Again in pursuing this relation there are advantages in working from a complete and totally bounded metric space.

8.34 Theorem. *A complete and totally bounded metric space* (X,d) *is ball cover compact.*

Proof. Suppose that (X,d) is not ball cover compact. Then there exists a cover for X by open balls $\{B_\alpha\}$ which has no finite subcover. So from the ε-net where $\varepsilon = \frac{1}{2}$ there exists an element x_1 such that there does not exist a finite subcover for $B(x_1;\frac{1}{2})$ from $\{B_\alpha\}$. (If each ball radius $\frac{1}{2}$ centred on a point of the ε-net did have a finite subcover from $\{B_\alpha\}$ then there would be a finite subcover for X.) But also from the ε-net where $\varepsilon = \frac{1}{4}$ there exists an element x_2, where $B(x_1;\frac{1}{2}) \cap B(x_2;\frac{1}{4}) \neq \phi$ such that there does not exist a finite subcover for $B(x_2;\frac{1}{4})$. (If each ball radius $\frac{1}{4}$ centred on a point of the ε-net and with non-empty intersection with $B(x_1;\frac{1}{2})$ did have a finite subcover from $\{B_\alpha\}$ then there would be a finite subcover of $B(x_1;\frac{1}{2})$.) Again from the ε-net where $\varepsilon = \frac{1}{8}$ there exists an element x_3, where $B(x_1;\frac{1}{2}) \cap B(x_2;\frac{1}{4}) \cap B(x_3;\frac{1}{8}) \neq \phi$, such that there does not exist a finite subcover for $B(x_3;\frac{1}{8})$.

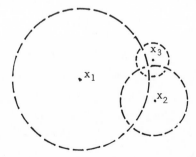

Figure 17.   The intersection of balls $\{B(x_n;\frac{1}{2^n})\}$ which does not have a finite subcover from $\{B_\alpha\}$.

Continuing inductively we have a sequence $\{x_n\}$ with the property that

$$d(x_{n-1}, x_n) \le \frac{1}{2^{n-1}} + \frac{1}{2^n} \le \frac{1}{2^{n-2}} \ .$$

Therefore, for $m > n$

$$d(x_n, x_m) \le d(x_n, x_{n+1}) + \ldots + d(x_{m-1}, x_m)$$

$$\le \frac{1}{2^{n-1}} + \ldots + \frac{1}{2^{m-2}} \le \frac{1}{2^{n-2}} \ ,$$

which implies that $\{x_n\}$ is a Cauchy sequence in (X,d). But since (X,d) is complete there exists an $x_0 \in X$ such that $\{x_n\}$ is convergent to x_0. We denote by B_{α_0} the open ball in the cover for X such that $x_0 \in B_{\alpha_0}$. Now there exists an $\varepsilon > 0$ such that $B(x_0; \varepsilon) \subseteq B_{\alpha_0}$. However, the sequence $\{x_n\}$ converges to x_0 so there exists a $\nu \in \mathbb{N}$ such that

$$d(x_\nu, x) < \frac{\varepsilon}{2} \text{ and } \frac{1}{2^\nu} < \frac{\varepsilon}{2} \ .$$

Then $B(x_\nu; \frac{1}{2^\nu}) \subseteq B(x_0; \varepsilon) \subseteq B_{\alpha_0} \ .$

But this contradicts the property that for $B(x_\nu; \frac{1}{2^\nu})$ there does not exist a finite subcover from $\{B_\alpha\}$.
We conclude that (X,d) is ball cover compact. $\square$

We now investigate how ball cover compactness relates to sequential compactness and as in Theorems 8.27 and 8.34 we work from a ball cover compact metric space.

<u>8.35 Theorem.</u> *A ball cover compact metric space* (X,d) *is sequentially compact.*

<u>Proof.</u> Consider a sequence $\{x_n\}$ in (X,d) and the associated set $A \equiv \{x_n : n \in \mathbb{N}\}$.
Suppose that A does not have a cluster point. Then from Theorem 4.7 we see that for each $x \in X$ there exists an $r(x) > 0$ such that $\big(B(x; r(x)) \backslash \{x\}\big) \cap A = \phi$. But the family of open balls $\big\{B\big((x; r(x)) : x \in X\big\}$ is an open ball cover for X. Since (X,d) is ball cover compact there exists a finite subcover.   But this implies that A is finite

and if so then the sequence $\{x_n\}$ has a subsequence convergent to one of its elements.

Suppose that A does have a cluster point. Then there exists a sequence in A convergent to the cluster point, and we conclude that the sequence $\{x_n\}$ has a convergent subsequence.

From both cases we conclude that (X,d) is sequentially compact. □

We have introduced three different conditions, sequential compactness, total boundedness and ball cover compactness and we have proved a chain of results which links to establish three equivalent forms of compactness for metric spaces.

8.36 Theorem. *For a metric space* (X,d) *the following conditions are equivalent,*

  (i)   (X,d) *is sequentially compact,*

  (ii)   (X,d) *is complete and totally bounded,*

  (iii) (X,d) *is ball cover compact.*

Proof.   Theorem 8.27 states that   (i) ⇒ (ii)

            Theorem 8.34 states that   (ii) ⇒ (iii) and

            Theorem 8.35 states that (iii) ⇒ (i).   □

8.37 Remark. We should notice that where sequential compactness is quite evidently an invariant for equivalent metrics, Theorem 8.36 implies that the equivalent conditions (ii) and (iii) are also invariant for equivalent metrics. This is an interesting observation especially for condition (ii) because Exercise 3.53.9 in the light of Corollary 8.29 shows that neither total boundedness nor completeness is separately invariant under equivalent metrics.   □

From this point on, when we are working with compactness we are at liberty to choose the form most suited to our purpose. In Section 9 we will establish key theorems which use all three forms and even in the same proof.

As a significant deduction from Theorem 8.36 we obtain a converse to Corollary 8.28 for complete metric spaces.

8.38 Corollary. *In a complete metric space a closed and totally bounded subset is compact.*

It is often convenient to use this result to establish compactness as we show in the following example.

8.39 Example. *In Hilbert sequence space* $(\ell_2, \|\cdot\|_2)$ *the Hilbert cube,* $C \equiv \{x \equiv \{\lambda_1, \lambda_2, \ldots, \lambda_n, \ldots\} \in \ell_2 : |\lambda_n| \le \frac{1}{n}$ *for all* $n \in \mathbb{N}\}$ *is compact.*

Proof. We first prove that C is totally bounded: Given $\varepsilon > 0$ there exists a $\nu \in \mathbb{N}$ such that $\sqrt{\left(\sum\limits_{\nu+1}^{\infty} \frac{1}{k^2}\right)} < \frac{\varepsilon}{2}$. Now consider the finite dimensional linear subspace ℓ_2^{ν} of ℓ_2 consisting of elements $x \equiv \{\lambda_1, \lambda_2, \ldots, \lambda_{\nu}, 0, \ldots\}$. Then $C_{\nu} \equiv C \cap \ell_2^{\nu}$ is a bounded subset of ℓ_2^{ν} and so is totally bounded. Therefore there exists an $\frac{\varepsilon}{2}$-net, $E \equiv \{x_1, x_2, \ldots, x_n\}$ in ℓ_2^{ν} such that

$$C_{\nu} \subseteq \cup\left\{B(x_k; \tfrac{\varepsilon}{2}) : k \in \{1, 2, \ldots, n\}\right\}.$$

But for any element $x \in C$, $d(x, C_{\nu}) < \frac{\varepsilon}{2}$.
So $C \subseteq \cup\left\{B(x_k; \varepsilon) : k \in \{1, 2, \ldots, n\}\right\}$;   that is, E is an ε-net for C in ℓ_2. Therefore E is totally bounded. It is not difficult to prove that C is also closed and so we conclude that C is compact in $(\ell_2, \|\cdot\|_2)$. $\square$

Compact metric spaces have   the special property of being separable. The form of compactness most appropriate to prove this property is total boundedness.

8.40 Theorem. *A totally bounded metric space* (X, d) *is separable.*

Proof. Since (X, d) is totally bounded, for each $n \in \mathbb{N}$ there exists an ε-net, E_n with $\varepsilon = \frac{1}{n}$. Now $E \equiv \cup\{E_n : n \in \mathbb{N}\}$ is countable. For any given $x \in X$, $d(x, E) \le d(x, E_n) < \frac{1}{n}$ for every $n \in \mathbb{N}$. So $d(x, E) = 0$ which implies that $\overline{E} = X$, (see Exercise 4.51.12(i)). $\square$

8.41 Exercises.

1.   In the spaces given determine whether the following subsets
are compact.

  (i)   In $\mathbb{R}$ with the usual norm,

     (a) $\{\frac{n}{n^2+1} : n \in \mathbb{Z}\}$,

     (b) $A \cap (-1,1)$ where A is the set of real numbers with
          decimal representation by digits 0 and 1.

  (ii)   In $\mathbb{C}$ with the usual norm,

     (a) $\{\frac{1}{z^2+1} : |z| = 1\}$,

     (b) given $r \in \mathbb{Q}$,

          $\{e^{nr\pi i} : n \in \mathbb{Z}\}$.

  (iii)   In $(\mathbb{R}^2, \|\cdot\|_2)$,

     (a) $\{(\lambda,\mu) : |\lambda| + |\mu| = 1\}$,

     (b) $\{(\lambda,\mu) : 0 \le \lambda\mu \le 1\}$.

  (iv)   In $(m, \|\cdot\|_\infty)$,

     (a) E the set of sequences whose terms consist only of digits
          0 and 1,

     (b) F the set of sequences

          $\{x \equiv \{\lambda_1,\lambda_2,\dots,\lambda_n,\dots\} : \lambda_n = 0, \frac{1}{n} \text{ or } -\frac{1}{n} \text{ for each } n \in \mathbb{N}\}$.

  (v)   In $(C[0,1], \|\cdot\|_\infty)$,

     (a) E the set of constant functions f where $\|f\|_\infty \le 1$,

     (b) $S(\underline{0};1)$ the unit sphere.

2.   Prove that a discrete metric space (X,d) is compact if and only if X is finite.

3.   (i)   Consider a sequence $\{x_n\}$ in a compact metric space (X,d). Prove that if the set $\{x_n : n \in \mathbb{N}\}$ has one and only one cluster point then the sequence $\{x_n\}$ is convergent.

    (ii)   Consider a Cauchy sequence $\{x_n\}$ in a metric space (X,d). Prove that the set $\{x_n : n \in \mathbb{N}\}$ is totally bounded.

4.   Prove that if $A_1, A_2, \ldots, A_n$ are compact subsets of a metric space then $\cup\{A_k : k \in \{1, 2, \ldots, n\}\}$ is compact.

5.   Prove that a metric space (X,d) is compact if and only if every infinite subset of X has a cluster point.

6.   (i)   Given a normed linear space $(X, \|\cdot\|)$, prove that if A is a compact subset and B is a closed subset then $A + B$ is closed.

    (ii)   Give an example in $\mathbb{R}$ with the usual norm of closed subsets A and B where $A + B$ is not closed.

7.   (i)   In a metric space (X,d), we are given a compact subset A and a closed subset B which are disjoint. Prove that $d(A,B) > 0$.

    (ii)   Give an example to show that this property does not always hold for disjoint closed subsets A and B.

8.   Prove that if T is a continuous mapping from a compact metric space (X,d) into a metric space (Y,d') then for every subset A of X we have $T(\overline{A}) = \overline{T(A)}$, (see Exercise 6.39.12).

9.   Consider a mapping T from a metric space (X,d) into a metric space (Y,d'). Prove that if the restriction of T to every compact subset of X is continuous, then T is continuous on X.
(Hint: Use Example 8.4.)

10. (i)   A mapping T from a metric space (X,d) into a metric space
           (Y,d') has the property that its graph is closed, (see
           Exercise 6.39.10(ii)). Prove that if (Y,d') is compact
           then T is continuous on X.

    (ii)   Consider a bounded real function f on a metric space (X,d).
           Prove that if f has a closed graph then f is continuous,
           (see Exercise 6.39.10(iii)).

11. (i)   A *compact operator* T on a normed linear space $(X, \|\cdot\|)$ is a
           linear operator with the property that $\overline{T(A)}$ is compact for
           every bounded subset A in $(X, \|\cdot\|)$. Prove that a compact
           operator is continuous.

    (ii)   A *finite rank operator* T on a normed linear space $(X, \|\cdot\|)$ is
           a linear operator such that $T(X)$ is finite dimensional.

           (a) Prove that every continuous finite rank operator is a
                compact operator.

           (b) Give an example of a finite rank operator which is not
                continuous.

12. (i)   For any metric space, prove that

           (a) a non-empty subset of a totally bounded set is totally
                bounded, and

           (b) the closure of a non-empty totally bounded set is totally
                bounded.

    (ii)   A subset A of a metric space (X,d) is said to be *relatively
           compact* if the closure $\overline{A}$ is compact. Prove that

           (a) in a metric space, a relatively compact set is totally
                bounded, and

           (b) in a complete metric space, a totally bounded set is
                relatively compact.

    (iii) Given a metric space (X,d), prove that a subset A is relatively
           compact if and only if every sequence $\{x_n\}$ in A has a
           subsequence convergent in X but with limit point not
           necessarily a member of A.

13.   Consider the Hilbert cube of Example 8.39,

$$C \equiv \{x \equiv \{\lambda_1, \lambda_2, \ldots, \lambda_n, \ldots\} \in \ell_2 : |\lambda_n| \leq \tfrac{1}{n} \text{ for all } n \in \mathbb{N}\} .$$

Determine whether

(i)   C is compact in $(\ell_2, \|\cdot\|_\infty)$,

(ii) $C \cap E_0$ is compact in $(E_0, \|\cdot\|_2)$.

14.   Given a subset A of a linear space X, the *convex hull* of A, denoted co A, is the set

$$\{ \sum_{k=1}^{n} \lambda_k a_k : n \in \mathbb{N} \text{ and set } \{a_1, a_2, \ldots, a_n\} \subseteq A$$
$$\text{and } \{\lambda_1, \lambda_2, \ldots, \lambda_n\} \subseteq \mathbb{R}^+ \text{ such that } \sum_{k=1}^{n} \lambda_k = 1\} .$$

Consider a normed linear space $(X, \|\cdot\|)$.

(i)   Prove that for a finite subset A, co A is compact.

(ii)   Prove that for a totally bounded subset A, co A is totally bounded.

(iii)   Prove that if $(X, \|\cdot\|)$ is complete, for a compact subset A, $\overline{\text{co A}}$ is compact.

(iv)   Consider the normed linear space $(E_0, \|\cdot\|_\infty)$ and the subset $A \equiv \{x_n : n \in \mathbb{N}\} \cup \{\underset{\sim}{0}\}$ where $x_n \equiv \{0, \ldots, 0, \underset{\text{nth place}}{\tfrac{1}{n}}, 0, \ldots\}$.

(a)   Prove that A is compact.

(b)   Show that the sequence $\{y_n\}$ where

$$y_n \equiv \sum_{k=1}^{n-1} \frac{x_k}{2^{k-1}} + \frac{x_n}{2^{n-1}}$$

is Cauchy in co A but not convergent in E_0.

(c)   Hence show that (iii) does not hold for incomplete normed linear spaces.

15. (i)   Prove that in a finite dimensional normed linear space the convex hull of every compact set is compact.

   (ii)   Prove that a normed linear space where the convex hull of every compact set is compact is finite dimensional.

16.   Let M be a finite dimensional linear subspace of a normed linear space $(X, \|\cdot\|)$. For the coset $[x] \in \frac{X}{M}$ prove that there exists an $x_0 \in [x]$ such that

$$\|x_0\| = \|[x]\|; \quad \text{(see Exercise 4.51.21)}.$$

17. (i)   Find a closed and bounded subset of $(m, \|\cdot\|_\infty)$ which has no point nearest to the origin.

   (ii)   In the real normed linear space $(C[0,1], \|\cdot\|_\infty)$ consider the linear subspace $C_0[0,1]$ of functions f where $f(0) = 0$, and the linear subspace $M[0,1]$ of $C_0[0,1]$ of functions f where $\int_0^1 f(t)dt = 0$. Show that there does not exist any $f \in C_0[0,1] \setminus M[0,1]$ where $\|f\|_\infty = 1$ and $d(f, M[0,1]) = 1$.

18.   (*A Fixed Point Theorem*)
Consider a mapping T of a metric space (X,d) into itself.

   (i)   Prove that the function $f : X \to \mathbb{R}$ defined by

   $$f(x) = d(Tx, x)$$

   is continuous on X.

   (ii)   Prove that if (X,d) is compact and T has the property

   $$d(Tx, Ty) < d(x,y) \quad \text{for all } x,y \in X, \ x \neq y,$$

   then T has a unique fixed point.

19.   Consider a mapping T of a compact metric space (X,d) into itself where T has the property

$$d(Tx,Ty) \geq d(x,y) \quad \text{for all } x,y \in X.$$

(i)   Prove that any given $x_0 \in X$ is a cluster point of the set of iterations $\{T^n x_0 : n \in \mathbb{N}\}$ and deduce that $T(X)$ is dense in X.

(ii)   Hence, or otherwise prove that T is an isometry of X onto X.

20.   Consider metric spaces (X,d) and (Y,d').

(i)   Prove that if A is a compact subset of X and B is a compact subset of Y then $A \times B$ is a compact subset of the product space $(X \times Y, d_\pi)$, (see Exercise 1.34.11).

(ii)   Prove that if C is a compact subset of $(X \times Y, d_\pi)$ then the set $\{x : (x,y) \in C\}$ is a compact subset of (X,d).

9. CONTINUOUS FUNCTIONS ON COMPACT METRIC SPACES

In real analysis, continuous functions defined on bounded closed intervals have specially useful properties. Continuous functions defined on a compact metric space have similar regular properties which distinguish them for applications.

Where it is not unduly inconvenient we will continue to work with sequential compactness although other forms of compactness may enable more direct argument.

We show firstly that there is a particularly simple criterion for a continuous mapping on a compact metric space to be a homeomorphism.

<u>9.1 Theorem</u>. *A continuous one-to-one mapping* T *from a compact metric space* (X,d) *onto a metric space* (Y,d') *is a homeomorphism.*

<u>Proof</u>. Suppose that T^{-1} is not continuous at $y_0 \in Y$. Then there exists an $r > 0$ and a sequence $\{y_n\}$ in Y such that $\{y_n\}$ is convergent to y_0 but the sequence $\{x_n\}$ where $x_n = T^{-1}y_n$ and $x_0 = T^{-1}y_0$ has the property that $d(x_n, x_0) > r$ for all $n \in \mathbb{N}$. However, since (X,d) is compact, $\{x_n\}$ has a subsequence $\{x_{n_k}\}$ which is convergent to some $x_1 \in X$. But T is continuous at x_1 so $\{y_{n_k}\}$ where $y_{n_k} = Tx_{n_k}$ is convergent to $y_1 = Tx_1$. But we are given that $\{y_{n_k}\}$ is convergent to y_0, so $y_1 = y_0$. Since T is one-to-one then $x_1 = x_0$ but this contradicts $d(x_{n_k}, x_0) > r$ for all $k \in \mathbb{N}$. $\square$

This theorem provides a standard method for establishing the continuity of an inverse function in real analysis.

<u>9.2 Example</u>. Consider $f : [0,\infty) \to \mathbb{R}$ defined by

$$f(t) = t^2.$$

Consider $t_0 \in [0,\infty)$ and $[0,a]$ where $0 \leqslant t_0 < a$. Now f is continuous and one-to-one on the compact set $[0,a]$, so by Theorem 9.1, f^{-1} is continuous on $[0,a^2]$. So f^{-1} is continuous at t_0^2 and therefore f^{-1} is continuous on $[0,\infty)$. $\square$

A particularly significant property of a continuous real function defined on a bounded closed interval is that it is uniformly continuous. However, this uniform continuity implication of compactness extends quite directly to the analysis of metric spaces.

9.3 Heine's Theorem. *A continuous mapping* T *from a compact metric space* (X,d) *into a metric space* (Y,d') *is uniformly continuous on* X.

<u>Proof.</u> Suppose that T is not uniformly continuous on X. Then there exists an r > 0 and sequences $\{x_n\}$ and $\{y_n\}$ in X such that $d(x_n,y_n) < \frac{1}{n}$ but

$$d'(Tx_n,Ty_n) \geqslant r \quad \text{for each } n \in \mathbb{N}.$$

Since (X,d) is compact the sequence $\{x_n\}$ has a subsequence $\{x_{n_k}\}$ convergent to some $x \in X$.
But for each $k \in \mathbb{N}$,

$$d\left(y_{n_k},x\right) \leqslant d\left(y_{n_k},x_{n_k}\right) + d\left(x_{n_k},x\right)$$

so the subsequence $\{y_{n_k}\}$ is also convergent to x. Since T is continuous at x, given $0 < \varepsilon < \frac{r}{2}$ there exists a $\nu \in \mathbb{N}$ such that

$$d'\left(Tx_{n_k},Tx\right) < \varepsilon \quad \text{for all } k > \nu.$$

However,

$$d'\left(Ty_{n_k},Tx\right) \geqslant d'\left(Ty_{n_k},Tx_{n_k}\right) - d'\left(Tx_{n_k},Tx\right)$$
$$> r - \varepsilon > \frac{r}{2} \quad \text{for all } k > \nu.$$

But this contradicts the continuity of T at x. $\square$

Naturally enough, in real analysis this theorem enables us to recognise uniform continuity very quickly.

9.4 Example. A continuous real function f on $[0,\infty)$ where $\lim_{t\to\infty} f(t) = a$ is uniformly continuous:

Since $\lim_{t\to\infty} f(t) = a$, given $\varepsilon > 0$ there exists a $K > 0$ such that so

$$\left|f(t) - a\right| < \frac{\varepsilon}{2} \qquad \text{when } t > K,$$

$$\left|f(t) - f(t')\right| < \varepsilon \qquad \text{for all } t, t' > K.$$

But from Heine's Theorem 9.3, f is uniformly continuous on $[0, K+1]$, so given $\varepsilon > 0$, there exists a $0 < \delta < 1$ such that

$$\left|f(t) - f(t')\right| < \varepsilon \qquad \text{for all } t, t' \in [0, \infty) \text{ when } |t-t'| < \delta. \quad \square$$

From Theorems 8.14 and 9.3 we can improve the statement of Theorem 9.1.

9.5 Corollary. *A continuous one-to-one mapping* T *from a compact metric space* (X,d) *onto a metric space* (Y,d') *is uniformly continuous on* X *and its inverse* T^{-1} *is uniformly continuous on* Y.

It has been apparent, since our extension of continuity to functions on metric spaces in Section 6 and especially as a consequence of Theorem 6.12 that, given a metric space (X,d), the set of bounded continuous scalar functions on X, denoted by $C(X)$, is a linear subspace of $B(X)$ and can be treated as a normed linear space with inherited supremum norm $\|\cdot\|_\infty$.

Furthermore, Theorem 3.14 can be directly extended for continuity of scalar functions on a metric space (X,d) to establish that the limit of a uniformly convergent sequence of continuous functions is itself continuous. As in Example 4.22, it follows that $(C(X), \|\cdot\|_\infty)$ is a Banach space.

We previously introduced $C(J)$ as the linear space of bounded continuous functions on the interval J and noted that for a bounded closed interval $[a,b]$, $C[a,b]$ is simply the linear space of all continuous functions on $[a,b]$. Similarly, when the metric space (X,d) is compact then we see from Theorem 8.14 that all continuous functions on X are bounded and $C(X)$ is the linear space of continuous functions on X.

Our analysis now focuses on the particular structural properties of the Banach space $(C(X), \|\cdot\|_\infty)$ when (X,d) is a compact metric space.

But firstly we prove an important uniform convergence result for continuous functions on a compact metric space. To do so we need a preliminary lemma which generalises Cantor's Intersection Theorem given in Exercise 4.51.20.

9.6 Lemma. *Given a compact metric space* (X,d), *if* $\{F_n\}$ *is a decreasing sequence of non-empty closed subsets then* $\cap\{F_n : n \in \mathbb{N}\}$ *is non-empty.*

Proof. For each $n \in \mathbb{N}$, choose $x_n \in F_n$. Since (X,d) is compact, the sequence $\{x_n\}$ has a subsequence $\{x_{n_k}\}$ which is convergent to $x \in X$. Since the sequence $\{F_n\}$ is decreasing, for each $n \in \mathbb{N}$ the subsequence $\{x_{n_k}\}$ lies in F_n for all but a finite number of terms. So either $x \in F_n$ or x is a cluster point of F_n. But F_n is closed so $x \in F_n$. But this holds for all $n \in \mathbb{N}$. Therefore $x \in \cap\{F_n : n \in \mathbb{N}\}$. $\square$

Our theorem on uniform convergence states that when (X,d) is a compact metric space, in the real Banach space $(C(X), \|\cdot\|_\infty)$, monotone and pointwise convergence imply uniform convergence.

9.7 Dini's Theorem. *Given a compact metric space* (X,d), *if* $\{f_n\}$ *is a monotone sequence in the real linear space* $C(X)$ *which is pointwise convergent to* $f \in C(X)$, *then* $\{f_n\}$ *is uniformly convergent to* f.

Proof. Suppose that $\{f_n\}$ is an increasing sequence. For each $n \in \mathbb{N}$, write $g_n \equiv f - f_n$. Then $\{g_n\}$ is a decreasing sequence of positive continuous functions pointwise convergent to 0. Given $\varepsilon > 0$, for each $n \in \mathbb{N}$ write

$$F_n \equiv \{x \in X : g_n(x) \geq \varepsilon\}.$$

Since each g_n is continuous on X it follows from Theorem 6.19 that each F_n is closed. Since $\{g_n\}$ is a decreasing sequence of functions, $\{F_n\}$ is a decreasing sequence of sets. Since $\{g_n\}$ is pointwise convergent to 0, $\cap\{F_n : n \in \mathbb{N}\} = \phi$. Therefore, from Lemma 9.6 there exists a $\nu \in \mathbb{N}$ such that $F_\nu = \phi$. So

$$0 \leq g_n(x) < \varepsilon \quad \text{for all } n > \nu \text{ and all } x \in X.$$

But then

$$\|f_n - f\|_\infty = \max\{|g_n(x)| : x \in X\}$$
$$< \varepsilon \quad \text{for all } n > \nu;$$

that is, $\{f_n\}$ is uniformly convergent to f on X. $\square$

<u>9.8 Remark</u>. We note that the continuity of the pointwise limit function is an important condition for Dini's Theorem. It is possible to have a monotone sequence of continuous functions pointwise convergent to a discontinuous limit function as in Example 3.13 but from Theorem 3.14 we see that such convergence cannot be uniform.

The compactness of the metric space (X,d) is also an important condition. If we modify Example 3.13 and consider the sequence $\{f_n\}$ on $[0,1)$ where $f_n(t) = t^n$, then we have a monotone sequence of continuous functions converging pointwise to a continuous function but convergence is not uniform. $\square$

9.9 The structure of the real Banach space $(C[a,b], \|\cdot\|_\infty)$

In real analysis it is a consequence of Heine's Theorem 9.3 that a continuous function f on a bounded closed interval $[a,b]$ can be approximated uniformly by step functions on $[a,b]$. In Example 4.4 we saw how such a function f on $[a,b]$ can be approximated uniformly by polygonal functions on $[a,b]$. We now show that f on $[a,b]$ can be approximated uniformly by polynomials on $[a,b]$. This result is known as Weierstrass' Approximation Theorem. By revealing that the polynomials are uniformly dense in the continuous functions on $[a,b]$, we gain insight into the nature of continuous functions on bounded closed intervals.

There are a number of proofs of the Weierstrass Approximation Theorem. The "constructive" proofs of this theorem, construct for any given continuous function f on $[a,b]$ and given $\varepsilon > 0$, a polynomial p on $[a,b]$ such that $\|f-p\|_\infty < \varepsilon$. The construction most commonly used in such proofs is that given by Bernstein and the approximating polynomials are called Bernstein polynomials. The construction due to Landau is perhaps closest to that originally proposed by Weierstrass.

We will follow an "existence" proof which establishes that such an approximating polynomial exists without actually constructing it.

On the face of it, an existence proof looks as though it gives less information and some would assert that philosophically such a proof is less satisfying. However, the advantages of the Bernstein and Landau constructions are illusory because in practice they do not generate a good algorithm for polynomial approximation. The main purpose of the construction is to establish existence. The advantage of the existence proof we give is that it generalises to reveal the structure of the Banach space $(C(X), \|\cdot\|_\infty)$ when (X,d) is a compact metric space.

9.9.1 The Weierstrass Approximation Theorem. *The linear subspace* $P[a,b]$ *of polynomials on* $[a,b]$ *is dense in the real Banach space* $(C[a,b], \|\cdot\|_\infty)$.

We prove this theorem through a sequence of lemmas.

9.9.2 Lemma. *There exists a sequence of polynomials* $\{p_n\}$ *on* $[-1,1]$ *which is increasing and uniformly convergent to the modulus function* $|\cdot|$ *on* $[-1,1]$.

Proof. We define the sequence $\{p_n\}$ on $[-1,1]$ by the following recurrence formula,

$$p_0(t) = 0 \quad \text{and}$$
$$p_{n+1}(t) = p_n(t) + \tfrac{1}{2}\left(t^2 - p_n^2(t)\right) \quad \text{for all } n \in \mathbb{N}.$$

Then $p_1(t) = \tfrac{1}{2}t^2 \leq |t| \quad$ for all $t \in [-1,1]$.

Now if $p_k(t) \leq |t|$ for some $k \in \mathbb{N}$ and all $t \in [-1,1]$ then

$$
\begin{aligned}
|t| - p_{k+1}(t) &= \left(|t| - p_k(t)\right) - \tfrac{1}{2}\left(t^2 - p_k^2(t)\right) \\
&= \left(|t| - p_k(t)\right)\left(1 - \tfrac{1}{2}(|t| + p_k(t))\right) \\
&\geq \left(|t| - p_k(t)\right)(1 - |t|) \\
&\geq 0 \quad \text{for all } t \in [-1,1].
\end{aligned}
$$

So by induction

$$p_n(t) \leq |t| \quad \text{for all } n \in \mathbb{N} \text{ and all } t \in [-1,1].$$

It is now clear from the recurrence formula that $\{p_n\}$ is an increasing sequence of positive functions bounded above by the modulus function $|\cdot|$. So by the Monotone Convergence Theorem of real analysis, the sequence $\{p_n\}$ is pointwise convergent to a function f on $[-1,1]$. Now f is a positive function and from the recurrence formula we deduce that

$$f(t) = f(t) + \tfrac{1}{2}\big(t^2 - f^2(t)\big) \quad \text{for all } t \in [-1,1]$$

from which we see that $f = |\cdot|$. Since the sequence $\{p_n\}$ is increasing and pointwise convergent to the continuous function $|\cdot|$ on $[-1,1]$, we have from Dini's Theorem 9.7 that $\{p_n\}$ is uniformly convergent to the modulus function $|\cdot|$ on $[-1,1]$. $\square$

9.9.3 Lemma. *In the Banach space* $(C[-1,1], \|\cdot\|_\infty)$,

(i)   *for every* $p \in P[-1,1]$ *we have* $|p| \in \overline{P[-1,1]}$,

(ii)   *for every* $f \in \overline{P[-1,1]}$ *we have* $|f| \in \overline{P[-1,1]}$.

Proof.   (i)   Given $\varepsilon > 0$, we have from Lemma 9.9.2 that there exists a sequence $\{p_n\}$ in $P[-1,1]$ and a $\nu \in \mathbb{N}$ such that

$$\big||t| - p_n(t)\big| < \varepsilon \quad \text{for all } n > \nu \text{ and all } t \in [-1,1].$$

Therefore, since

$$\frac{|p(t)|}{\|p\|_\infty} \leqslant 1 \quad \text{for all } t \in [-1,1],$$

we have $\left| \dfrac{|p(t)|}{\|p\|_\infty} - p_n\!\left(\dfrac{p(t)}{\|p\|_\infty}\right) \right| < \varepsilon \quad$ for all $n > \nu$ and all $t \in [-1,1]$.

Now for each $n \in \mathbb{N}$, $p_n\!\left(\dfrac{p(t)}{\|p\|_\infty}\right)$ is a polynomial on $[-1,1]$, so we conclude that $|p| \in \overline{P[-1,1]}$.

(ii)   Since $f \in \overline{P[-1,1]}$, given $\varepsilon > 0$ we have from Remark 4.26(ii) that there exists a $p \in P[-1,1]$ such that $\|f-p\|_\infty < \varepsilon$ and so

$$|f(t) - p(t)| < \varepsilon \quad \text{for all } t \in [-1,1].$$

But then

$$||f(t)|-|p(t)|| < \varepsilon \quad \text{for all } t \in [-1,1]$$

and so $|| |f|-|p| ||_\infty < \varepsilon$.   Again from Remark 4.26(ii) and Theorem 4.28(i) we conclude that $|f| \in \overline{P[-1,1]}$. $\square$

9.9.4 Definition. For real functions f and g on a set X, the real functions $f \vee g$ and $f \wedge g$ on X are defined by

$$f \vee g(x) = \max\{f(x),g(x)\}$$

$$f \wedge g(x) = \min\{f(x),g(x)\}.$$

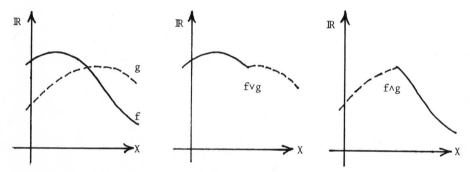

Figure 18.   Graphical representation of f∨g and f∧g.

9.9.5 Remark. For real functions f and g on a set X,

$$f \vee g = \tfrac{1}{2}(f+g+|f-g|)$$

$$f \wedge g = \tfrac{1}{2}(f+g-|f-g|).$$

It is clear that for continuous functions f and g on a metric space (X,d), $f \vee g$ and $f \wedge g$ are also continuous functions on (X,d). $\square$

Using Remark 9.9.5 and Theorem 4.33 the following result is immediate.

9.9.6 Corollary. *In the Banach space* $(C[-1,1], \|\cdot\|_\infty)$, *for* $f,g \in \overline{P[-1,1]}$ *we have* $f \vee g, f \wedge g \in \overline{P[-1,1]}$.

We are now in position to tackle the main step in our proof.

9.9.7 Lemma. *The linear subspace* $P[-1,1]$ *is dense in the Banach space* $(C[-1,1], \|\cdot\|_\infty)$.

Proof. Consider $f \in C[-1,1]$ and $x,y \in [-1,1]$. Now there exists a $p_{xy} \in P[-1,1]$ such that

$$p_{xy}(x) = f(x) \quad \text{and} \quad p_{xy}(y) = f(y).$$

Given $\varepsilon > 0$ and $x \in [-1,1]$ for each $y \in [-1,1]$ consider a polynomial p_{xy}. Since f and p_{xy} are continuous at y and $p_{xy}(y) = f(y)$, there exists a $\delta(y) > 0$ such that

$$|p_{xy}(t) - f(t)| < \varepsilon \quad \text{for} \quad |t-y| < \delta(y).$$

The family of intervals $\{(y-\delta(y), y+\delta(y)) : y \in [-1,1]\}$ is a cover for $[-1,1]$ by open balls. But since $[-1,1]$ is compact there exists a finite subcover

$$\{(y_k - \delta(y_k), y_k + \delta(y_k)) : k \in \{1,2,\ldots,n\}\} \quad \text{for} \quad [-1,1].$$

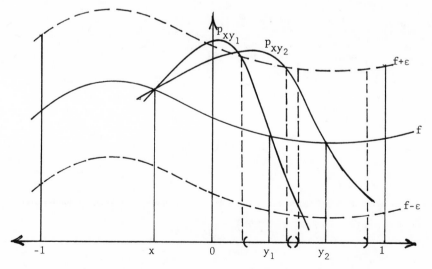

Figure 19.   Polynomials p_{xy_1} and p_{xy_2}.

We define the function g_x on $[-1,1]$ by

$$g_x = \vee \{ p_{xy_k} : k \in \{1,2,\ldots,n\} \}.$$

By Corollary 9.9.6 we have that $g_x \in \overline{P[-1,1]}$. For each $t \in [-1,1]$ there exists a $k \in \{1,2,\ldots,n\}$ such that $t \in \left(y_k - \delta(y_k), y_k + \delta(y_k) \right)$ and

$$f(t) < p_{xy_k}(t) + \varepsilon$$

and so

$$f(t) < g_x(t) + \varepsilon \quad \text{for all } t \in [-1,1].$$

Now for each $x \in [-1,1]$ consider g_x. Since f and g_x are continuous at x and $g_x(x) = f(x)$ there exists a $\delta(x) > 0$ such that

$$\left| g_x(t) - f(t) \right| < \varepsilon \quad \text{for } |t-x| < \delta(x).$$

The family of intervals $\left\{ \left(x - \delta(x), x + \delta(x) \right) : x \in [-1,1] \right\}$ is a cover for $[-1,1]$ by open balls. But again since $[-1,1]$ is compact there exists a finite subcover

$$\left\{ \left(x_k - \delta(x_k), x_k + \delta(x_k) \right) : k \in \{1,2,\ldots,m\} \right\} \quad \text{for } [-1,1].$$

We define the function g on $[-1,1]$ by

$$g = \wedge \{ g_{x_k} : k \in \{1,2,\ldots,m\} \}.$$

By Corollary 9.9.6 we have that $g \in \overline{P[-1,1]}$. For each $t \in [-1,1]$ there exists a $k \in \{1,2,\ldots,m\}$ such that $t \in \left(x_k - \delta(x_k), x_k + \delta(x_k) \right)$ and

$$f(t) > g_{x_k}(t) - \varepsilon$$

and so

$$f(t) > g(t) - \varepsilon \quad \text{for all } t \in [-1,1].$$

But we also have for every x $\in$ [-1,1] that

$$f(t) < g_x(t) + \varepsilon \quad \text{for all } t \in [-1,1]$$

and so

$$f(t) < g(t) + \varepsilon \quad \text{for all } t \in [-1,1].$$

Therefore, $\|f-g\|_\infty < \varepsilon$. But $g \in \overline{P[-1,1]}$ so from Theorem 4.33 we conclude that $f \in \overline{P[-1,1]}$. $\square$

The final step in the proof scales the result in Lemma 9.9.7 from the interval [-1,1] to the interval [a,b].

Proof of the Weierstrass Approximation Theorem 9.9.1. For any continuous function f on [a,b] we define the function g on [-1,1] by

$$g(t) = f\left(\tfrac{1}{2}(a+b)+(b-a)t\right).$$

Then g is continuous on [-1,1] and by Lemma 9.9.7, given $\varepsilon > 0$ there exists a polynomial q on [-1,1] such that $\|q-p\|_\infty < \varepsilon$. We define the function p on [a,b] by

$$p(t) = q\left(\frac{2t-(a+b)}{b-a}\right).$$

Then p is a polynomial on [a,b] and

$$\|f-p\|_\infty = \|g-q\|_\infty < \varepsilon. \;\square$$
$$\text{on } [a,b] \quad \text{on } [0,1]$$

From the Weierstrass Approximation Theorem 9.9.1 we can deduce the following structural property of the Banach space $(C[a,b], \|\cdot\|_\infty)$.

9.9.8 Corollary. *The real Banach space* $(C[a,b], \|\cdot\|_\infty)$ *is separable.*

Proof. Consider the linear space $P[a,b]$ of polynomials on [a,b] and the subset $P_{\mathbb{Q}}[a,b]$ of polynomials on [a,b] with rational coefficients. For any given polynomial p on [a,b] where

$$p(t) \equiv a_0 + a_1 t + \ldots + a_m t^m$$

we see from the proof of Theorem 4.43 that there exists a sequence of mth degree polynomials $\{q_n\}$ in $P_Q[a,b]$ convergent to p. So $P_Q[a,b]$ is dense in $(P[a,b], \|\cdot\|_\infty)$. From the Weierstrass Approximation Theorem 9.9.1 and Theorem 4.33, $P_Q[a,b]$ is dense in $(C[a,b], \|\cdot\|_\infty)$. Since Q^m is countable for every $m \in \mathbb{N}$ we have that $P_Q[a,b]$ is countable. We conclude that $P_Q[a,b]$ is a countable dense subset of $(C[a,b], \|\cdot\|_\infty)$. $\Box$

9.9.9 Remark. In real analysis, Taylor's Theorem provides a technique for uniformly approximating many infinitely often differentiable functions by Taylor polynomials. We recall that this type of polynomial approximation is extremely useful because the approximation is improved by the addition of higher degree terms in the polynomial. In the Weierstrass Approximation Theorem there is no implicit implication even in the constructive proofs that better polynomial approximations are necessarily gained in this way. It is for this reason that we can say that a "constructive" proof has no practical advantage over the "existence" proof. $\Box$

9.10 The structure of the Banach space $(C(X), \|\cdot\|_\infty)$ where (X,d) is a compact metric space.

We did claim some advantage for the "existence" proof in that it could be applied to a more general situation.

9.10.1 Definition. A set L of real functions on a set X is called a *lattice* if $f \vee g$, $f \wedge g \in L$ for all $f,g \in L$.

Notice that Lemmas 9.9.2, 9.9.3 and Corollary 9.9.6 simply establish that $\overline{P[-1,1]}$ is a lattice in $(C[-1,1], \|\cdot\|_\infty)$ and Lemma 9.9.7 simply used this lattice property of $\overline{P[-1,1]}$. Consequently it is not difficult to generalise Lemma 9.9.7 to produce an even more significant result. The framework for this generalisation is contained in the following theorem.

9.10.2 Stone's Approximation Theorem. *Given a compact metric space* (X,d), *if*

(i)   L *is a lattice in the real linear space* $C(X)$ *and*

(ii)   *for every* $a,b \in \mathbb{R}$ *and* $x,y \in X$, $x \neq y$, *there exists an* $f \in L$
       such that $f(x) = a$ *and* $f(y) = b$

then L *is dense in* $(C(X), \|\cdot\|_\infty)$.

It should be noticed that $P[-1,1]$ is a subalgebra of $C[-1,1]$; (see Exercise 4.51.17(i)). We use the structure of $C(X)$ as an algebra to generalise Lemma 9.9.3 and Corollary 9.9.6.

9.10.3 Lemma. *Given a metric space* (X,d), *if* A *is a closed subalgebra of the real Banach space* $(C(X), \|\cdot\|_\infty)$ *then* A *is a lattice.*

Proof. For any $f \in A$,

$$\frac{|f(x)|}{\|f\|_\infty} \leq 1 \quad \text{for all } x \in X.$$

Using the sequence of polynomials $\{p_n\}$ given by the recurrence formula in Lemma 9.9.2 we have as in Lemma 9.9.3 that, given $\varepsilon > 0$ there exists a $\nu \in \mathbb{N}$ such that

$$\left| \frac{|f(x)|}{\|f\|_\infty} - p_n\left(\frac{f(x)}{\|f\|_\infty}\right) \right| < \varepsilon \quad \text{for all } n > \nu \text{ and all } x \in X.$$

Again from the recurrence formula we see that $p_n(0) = 0$ for all $n \in \mathbb{N}$ so $p_n \circ \dfrac{f}{\|f\|_\infty} \in A$ and therefore $|f| \in A$. But then using the formula of Remark 9.9.5 we see that $f \vee g, f \wedge g \in A$ for all $f,g \in A$. $\square$

We are now in a position to give the generalisation of the Weierstrass Approximation Theorem.

9.10.4 The Stone-Weierstrass Theorem. *Given a compact metric space* (X,d), *if* A *is a subalgebra of the real linear space* $C(X)$ *where*

(i)   *the constant function* $1 \in A$, *and*

(ii)   *for every* $x,y \in X$, $x \neq y$, *there exists an* $f \in A$ *such that*
       $f(x) \neq f(y)$,

then A *is dense in* $(C(X), \|\cdot\|_\infty)$.

<u>Proof</u>. Consider $a,b \in \mathbb{R}$ and $x,y \in X$, $x \neq y$. From (ii) there exists an
$f \in A$ such that $f(x) \neq f(y)$ so it follows that there exist $\alpha, \beta \in \mathbb{R}$ such

that $\alpha f(x) + \beta = a$

and $\alpha f(y) + \beta = b$.

The function $g \equiv \alpha f + \beta \in A$ and $g(x) = a$ and $g(y) = b$. Now $\overline{A}$ is also a
subalgebra, (see Exercise 4.51.17(ii)), so by Lemma 9.10.3, $\overline{A}$ is a lattice.
Naturally $\overline{A}$ also has properties (i) and (ii) of the theorem statement.
It now follows from Stone's Theorem 9.10.2 that $\overline{A}$ is dense in $(C(X), \|\cdot\|_\infty)$
and so $\overline{A} = C(X)$; that is, A is dense in $(C(X), \|\cdot\|_\infty)$. $\square$

Now the Stone-Weierstrass Theorem does not hold in general for
a complex Banach space $(C(X), \|\cdot\|_\infty)$; nevertheless, under slightly modified
conditions the theorem can be extended to the complex case.

<u>9.10.5 Definition</u>. Given a continuous complex function f on a metric space
(X,d) the functions $\overline{f}$, Re f and Im f are defined on X by

$$\overline{f}(x) = \overline{f(x)},$$

$$(\text{Re } f)(x) = \text{Re } f(x) \quad \text{and}$$

$$(\text{Im } f)(x) = \text{Im } f(x).$$

<u>9.10.6 Remark</u>. Now $|\overline{f}(x) - \overline{f}(y)| = |f(x) - f(y)|$ for all $x, y \in X$ so $\overline{f}$ is
continuous on X. But also

$$\text{Re } f = \tfrac{1}{2}(f + \overline{f}), \quad \text{Im } f = \frac{1}{2i}(f - \overline{f})$$

so Re f and Im f are continuous real functions on X. $\square$

To extend the Stone-Weierstrass Theorem we restrict our
attention to those subalgebras A of the complex linear space $C(X)$ where
if $f \in A$ then $\overline{f} \in A$.

9.10.7 The Complex Stone-Weierstrass Theorem. *Given a compact metric space* (X,d), *if* A *is a subalgebra of the complex linear space* $C(X)$ *where*

(i)   *the constant function* $1 \in A$,

(ii)   *for every* $x,y \in X$, $x \neq y$, *there exists an* $f \in A$ *such that* $f(x) \neq f(y)$, *and*

(iii) $\overline{f} \in A$ *whenever* $f \in A$

then A *is dense in* $(C(X), \|\cdot\|_\infty)$.

Proof. Now $\overline{A}$ is a closed subalgebra of $(C(X), \|\cdot\|_\infty)$, (see Exercise 4.51.17(ii)), so

$$B \equiv \overline{A} \cap \text{real } C(X)$$

is a closed subalgebra of the real Banach space $(C(X), \|\cdot\|_\infty)$. For each $f \in A$ we have $\overline{f} \in A$ so from Remark 9.10.6 we have that Re f, Im $f \in A$, so

$$\text{Re } f, \text{Im } f \in A \cap \text{real } C(X) \subseteq B.$$

Given $x,y \in X$, $x \neq y$ there exists an $f \in A$ such that $f(x) \neq f(y)$, so

$$\text{Re } f(x) \neq \text{Re } f(y) \quad \text{or} \quad \text{Im } f(x) \neq \text{Im } f(y)$$

which implies that there exists a $g \in B$ such that $g(x) \neq g(y)$. From the Stone-Weierstrass Theorem 9.10.4 we conclude that $B = \text{real } C(X)$. But then for each $f \in$ complex $C(X)$ we have

$$\text{Re } f, \text{Im } f \in B \subseteq \overline{A}$$

so $f = \text{Re } f + i \text{ Im } f \in \overline{A}$; that is, $\overline{A} = C(X)$. $\square$

In Section 9.9 we proved the Weierstrass Approximation Theorem only for the real Banach space $(C[a,b], \|\cdot\|_\infty)$ but it is clear that given any complex polynomial p on $[a,b]$ then $\overline{p}$ is also a complex polynomial on $[a,b]$ and so $P[a,b]$ in the complex Banach space $(C[a,b], \|\cdot\|_\infty)$ satisfies the subalgebra conditions (i), (ii) and (iii) of Theorem 9.10.7.

<u>9.10.8 Corollary</u>. *The linear space* P[a,b] *of polynomials on* [a.b] *is dense in the complex Banach space* $(C[a,b], \|\cdot\|_\infty)$.

We now deduce from the Stone-Weierstrass Theorem 9.10.4 that the structural property proved for $(C[a,b], \|\cdot\|_\infty)$ in Corollary 9.9.8 holds in general.

<u>9.10.9 Theorem</u>. *Given a compact metric space* (X,d), *the Banach space* $(C(X), \|\cdot\|_\infty)$ *is separable.*

<u>Proof.</u>   (i) Consider firstly the real linear space $C(X)$. Now by Theorem 8.40, the compact metric space (X,d) is separable; that is, there exists a countable set $\{x_n : n \in \mathbb{N}\}$ dense in (X,d). For each $n \in \mathbb{N}$ define the real function f_n on X by

$$f_n(x) = d(x, x_n).$$

Now each f_n is continuous on X so $f_n \in C(X)$ for all $n \in \mathbb{N}$. Consider A the subalgebra generated by $\{f_n : n \in \mathbb{N}\}$ and the constant function 1. For every $x,y \in X$, $x \neq y$ it is clear that there exists an $x_n \in X$ such that $d(x,x_n) < d(y,x_n)$ so $f_n(x) \neq f_n(y)$. Therefore A satisfies the conditions of the Stone-Weierstrass Theorem 9.10.4 and so A is dense in $(C(X), \|\cdot\|_\infty)$.

Now consider $A_\mathbb{Q}$ the subalgebra generated by $\{f_n : n \in \mathbb{N}\}$ and the constant function 1 over the rationals $\mathbb{Q}$. Now A consists of the linear span of products of the form $f_1^{m_1}, f_2^{m_2} \ldots f_n^{m_n}$ for $n \in \mathbb{N}$ and $m_k \in \mathbb{N}$ for each $k \in \{1,2,\ldots,n\}$. Now the set of such products is countable so $A_\mathbb{Q}$ is countable. As in Corollary 9.9.8 we see that $A_\mathbb{Q}$ is dense in $(A, \|\cdot\|_\infty)$.

We conclude from Theorem 4.33 that $A_\mathbb{Q}$ is dense in $(C(X), \|\cdot\|_\infty)$ and so real $(C(X), \|\cdot\|_\infty)$ is separable.

        (ii) Consider now the complex linear space $C(X)$. If $\{g_n : n \in \mathbb{N}\}$ is a countable dense subset in real $C(X)$, the subset $\{g_n + ig_m : m,n \in \mathbb{N}\}$ is a countable subset of complex $C(X)$. For $f \in C(X)$ we have Re f, Im f $\in$ real $C(X)$ so given $\varepsilon > 0$ there exist $m,n \in \mathbb{N}$ such that

$$\|\text{Re } f - g_n\|_\infty < \varepsilon$$

and $$\|\text{Im } f - g_m\|_\infty < \varepsilon.$$

Therefore,

$$\|f-(g_n + ig_m)\|_\infty \leq \|Re\ f - g_n\|_\infty + \|Im\ f - g_m\|_\infty$$

$$< 2\varepsilon$$

so by Remark 4.35, the subset

$$\{g_n + ig_m : m,n \in \mathbb{N}\} \text{ is dense in } (C(X),\|\cdot\|_\infty)$$

and so complex $(C(X),\|\cdot\|_\infty)$ is separable. $\square$

The classical Weierstrass Approximation Theorem by trigonometric polynomials follows as a corollary of the Stone-Weierstrass Theorem 9.10.4.

9.10.10 Definition. A *trigonometric polynomial* p is real function on $\mathbb{R}$ of the form

$$p(\theta) = \tfrac{1}{2}\alpha_0 + \sum_{k=1}^{n} (\alpha_k \cos k\theta + \beta_k \sin k\theta)$$

where $\alpha_k,\beta_k \in \mathbb{R}$ for all $k \in \{0,1,2,\ldots,n\}$.
A trigonometric polynomial is clearly periodic with period 2π.

9.10.11 The Weierstrass Approximation Theorem for trigonometric polynomials.
The linear subspace of trigonometric polynomials on $\mathbb{R}$ is uniformly dense in the linear space $\tilde{C}(2\pi)$ of continuous real periodic functions of period 2π on $\mathbb{R}$.

Proof. Consider the unit circle $\Gamma \equiv \{(\lambda,\mu) : \lambda^2+\mu^2 = 1\}$ in $(\mathbb{R}^2,\|\cdot\|_2)$. The linear space $\tilde{C}(2\pi)$ is isomorphic to real $C(\Gamma)$ under the mapping $f \mapsto f^*$ defined by

$$f^*(\cos \theta, \sin \theta) = f(\theta) \quad \text{for all } \theta \in \mathbb{R}$$

and this is an isometric isomorphism when the supremum norm is assumed on each linear space. Now by Theorem 8.9, Γ is compact in $(\mathbb{R}^2,\|\cdot\|_2)$. The set A*, the image of the linear subspace of trigonometric polynomials on $\mathbb{R}$ under the isomorphism is a subalgebra of $C(\Gamma)$ which contains the constant

function 1 and for every x,y $\in$ Γ, x $\neq$ y, there exists a p* $\in$ A* such that
p*(x) $\neq$ p*(y). So by the Stone-Weierstrass Theorem 9.10.4, A* is dense in
$(C(\Gamma), \|\cdot\|_\infty)$ and the theorem conclusion is a consequence of the isometric
isomorphism. $\square$

9.11 Compactness in $(C(X), \|\cdot\|_\infty)$.

From Theorem 8.36 we see that a closed subset of a complete
metric space is compact if and only if it is totally bounded. Compact
subsets in the particular spaces $(C(X), \|\cdot\|_\infty)$ where (X,d) is a compact
metric space have a characterisation which is somewhat simpler to apply.
In this case total boundedness is replaced by boundedness and a certain
continuity criterion affecting the set of functions in the subset.

9.11.1 Definition. Given a metric space (X,d) a non-empty subset A of
$C(X)$ is said to be *equicontinuous* at x_0 $\in$ X if given ϵ > 0 there exists a
$\delta(\epsilon, x_0)$ > 0 such that

$$\left| f(x) - f(x_0) \right| < \epsilon \quad \text{when} \quad d(x, x_0) < \delta \quad \text{for all } f \in A.$$

The subset A is said to be *equicontinuous on* X if A is equicontinuous at
each x $\in$ X.
The subset A is said to be *uniformly equicontinuous* on X if given ϵ > 0
there exists a $\delta(\epsilon)$ > 0 such that for all x,y $\in$ X

$$\left| f(x) - f(y) \right| < \epsilon \quad \text{when } d(x,y) < \delta \quad \text{for all } f \in A.$$

9.11.2 Remark. Notice that in the definition of equicontinuity δ depends
on ϵ and x_0 but is independent of any particular f $\in$ A. Clearly if A is a
finite subset then A is always equicontinuous at every x $\in$ X. More
generally any finite union of equicontinuous sets in $C(X)$ is an equicontinuous set.

Notice that in the definition of uniform equicontinuity δ
depends only on ϵ but is independent of any particular x $\in$ X or f $\in$ A.
Of course each function f in a uniformly equicontinuous set is uniformly
continuous.

In the special case when (X,d) is a compact metric space, we
see from Heine's Theorem 9.3 that every continuous function on X is
uniformly continuous on X. So an equicontinuous subset of $C(X)$ where
(X,d) is a compact metric space is uniformly equicontinuous. $\square$

We begin by exploring the implications of equicontinuity.

9.11.3 Lemma. *Given a metric space* (X,d), *if a sequence* $\{f_n\}$ *in* $B(X)$ *is equicontinuous at* x_0 *and is pointwise convergent to* f *on* X *then* f *is continuous at* x_0.

Proof. Given $\varepsilon > 0$ there exists a $\delta(\varepsilon, x_0) > 0$ such that

$$\left| f_n(x) - f_n(x_0) \right| < \varepsilon \quad \text{when } d(x,x_0) < \delta \quad \text{for all } n \in \mathbb{N}.$$

But since $\{f_n\}$ is pointwise convergent to f on X,

$$\left| f(x) - f(x_0) \right| \leq \varepsilon \quad \text{when } d(x,x_0) < \delta;$$

that is, f is continuous at x_0. $\square$

9.11.4 Remark. It is clear that the sequence $\{f_n\}$ in $B[0,1]$ where $f_n(t) = t^n$ is not equicontinuous on $[0,1]$ because the sequence $\{f_n\}$ has a discontinuous pointwise limit function;   (see Example 3.13). $\square$

9.11.5 Lemma. *Given a metric space* (X,d), *if a sequence* $\{f_n\}$ *in* $C(X)$ *is equicontinuous on* X *and is pointwise convergent on a dense subset* D *of* X *then* $\{f_n\}$ *is pointwise convergent on* X.

Proof. Consider a point $x \in X$. Now given $\varepsilon > 0$ there exists a $\delta(\varepsilon, x) > 0$ such that

$$\left| f_n(x) - f_n(y) \right| < \varepsilon \quad \text{when } d(x,y) < \delta \quad \text{for all } n \in \mathbb{N}.$$

Since D is dense in (X,d) there exists a $y \in D$ such that $d(x,y) < \delta$. Now $\{f_n(y)\}$ is convergent so is a Cauchy sequence of scalars, that is, there exists a $\nu \in \mathbb{N}$ such that

$$\left| f_n(y) - f_m(y) \right| < \varepsilon \quad \text{when } m,n > \nu.$$

So
$$\left| f_n(x) - f_m(x) \right| \leq \left| f_n(x) - f_n(y) \right| + \left| f_n(y) - f_m(y) \right| + \left| f_m(y) - f_m(x) \right|$$
$$< 3\varepsilon \quad \text{for all } m,n > \nu;$$

that is, $\{f_n(x)\}$ is a Cauchy sequence of scalars and so is convergent. $\square$

9.11.6 Lemma. *Given a compact metric space* (X,d), *if a sequence* $\{f_n\}$ *in* $C(X)$ *is equicontinuous on* X *and is pointwise convergent to* f *on* X *then* $\{f_n\}$ *is uniformly convergent to* f.

Proof. Since (X,d) is compact, the sequence $\{f_n\}$ is uniformly equicontinuous on X. So given $\varepsilon > 0$ there exists a $\delta(\varepsilon) > 0$ such that for all $x,y \in X$.

$$|f_n(x) - f_n(y)| < \varepsilon \quad \text{when } d(x,y) < \delta \text{ for all } n \in \mathbb{N}.$$

The family $\{B(x;\delta) : x \in X\}$ is a cover for X by open balls. So again since (X,d) is compact there exists a finite subcover $\{B(x_k;\delta) : k \in \{1,2,\ldots,m\}\}$ for X. Now $\{f_n\}$ is pointwise convergent to f on X. So for each $k \in \{1,2,\ldots,m\}$ there exists a $\nu_k \in \mathbb{N}$ such that

$$|f_n(x_k) - f(x_k)| < \varepsilon \quad \text{for all } n > \nu_k.$$

For any $x \in X$ there exists a $k_0 \in \{1,2,\ldots,m\}$ such that $d(x,x_{k_0}) < \delta$ and so

$$|f_n(x) - f(x)| \leq |f_n(x) - f_n(x_{k_0})| + |f_n(x_{k_0}) - f(x_{k_0})| + |f(x_{k_0}) - f(x)|$$
$$< 3\varepsilon \quad \text{for all } n > \nu_{k_0}$$

But then for all $x \in X$,

$$|f_n(x) - f(x)| < 3\varepsilon \quad \text{for all } n > \nu \equiv \max\{\nu_k : k \in \{1,2,\ldots,m\}\}.$$

Therefore $\|f_n - f\|_\infty \leq 3\varepsilon$ for all $n > \nu$; that is, $\{f_n\}$ is uniformly convergent to f. $\square$

We are now ready to prove the theorem characterising compact subsets of $(C(X), \|\cdot\|_\infty)$ when (X,d) is a compact metric space.

9.11.7 The Ascoli-Arzelà Theorem.

Given a compact metric space (X,d), *a closed subset* A *of the Banach space* $(C(X), \|\cdot\|_\infty)$ *is compact if and only if* A *is bounded and equicontinuous on* X.

Proof. Suppose that A is compact. By Theorem 8.7(i) A is bounded. We show that A is equicontinuous on X:

From Theorem 8.36, A is totally bounded; that is, given $\varepsilon > 0$ there exists an ε-net $\{f_1, f_2, \ldots, f_m\}$ in A. Now by Heine's Theorem 9.3, for each $k \in \{1, 2, \ldots, m\}$, f_k is uniformly continuous on X so there exists a $\delta_k > 0$ such that for all $x, y \in X$

$$|f_k(x) - f_k(y)| < \varepsilon \quad \text{when } d(x,y) < \delta_k.$$

Now for each $f \in A$ there exists a $k_0 \in \{1, 2, \ldots, m\}$ such that

$$\|f - f_{k_0}\|_\infty < \varepsilon.$$

Therefore, for each $f \in A$ and all $x, y \in X$

$$\begin{aligned}
|f(x) - f(y)| &\leq |f(x) - f_{k_0}(x)| + |f_{k_0}(x) - f_{k_0}(y)| + |f_{k_0}(y) - f(y)| \\
&< 2\|f - f_{k_0}\|_\infty + |f_{k_0}(x) - f_{k_0}(y)| \\
&< 3\varepsilon \quad \text{when } d(x,y) < \delta \equiv \min\{\delta_1, \delta_2, \ldots, \delta_m\};
\end{aligned}$$

that is, A is equicontinuous on X.

Conversely, suppose that A is bounded and equicontinuous on X. We show that A is sequentially compact:

Since (X,d) is compact we have from Theorem 8.40 that (X,d) is separable; that is, there exists a countable subset $\{x_n : n \in \mathbb{N}\}$ dense in (X,d). Consider any sequence $\{f_n\}$ in A. Since A is bounded, the sequence $\{f_n\}$ is bounded and so for each $x \in X$, the sequence $\{f_n(x)\}$ is a bounded sequence of scalars and so has a convergent subsequence. Let $\{f_{n1}\}$ denote the subsequence of $\{f_n\}$ such that $\{f_{n1}(x_1)\}$ is convergent and let $\{f_{n2}\}$ denote the subsequence of $\{f_{n1}\}$ such that $\{f_{n2}(x_2)\}$ is convergent. Continuing this process we have by induction a sequence of subsequences of the form

$$\{f_n\} \equiv \{f_1, f_2, \ldots, f_n, \ldots\}$$
$$\{f_{n1}\} \equiv \{f_{11}, f_{21}, \ldots, f_{n1}, \ldots\}$$
$$\{f_{n2}\} \equiv \{f_{12}, f_{22}, \ldots, f_{n2}, \ldots\}$$
$$\cdots$$
$$\{f_{nk}\} \equiv \{f_{1k}, f_{2k}, \ldots, f_{nk}, \ldots\}$$
$$\cdots$$

where each is a subsequence of the one above and $\{f_{nk}(x_k)\}$ is convergent for each $k \in \mathbb{N}$. Consider the diagonal subsequence $\{f_{nn}\}$ where

$$\{f_{nn}\} \equiv \{f_{11}, f_{22}, \ldots, f_{nn}, \ldots\}$$

Now this subsequence is pointwise convergent on $\{x_n : n \in \mathbb{N}\}$. By Lemma 9.11.5, $\{f_{nn}\}$ is pointwise convergent to some f on X. But also from Lemma 9.11.3, f is continuous on X. Further by Lemma 9.11.6 the subsequence $\{f_{nn}\}$ is uniformly convergent to f. But A is closed so $f \in A$ and we conclude that A is sequentially compact. □

The Ascoli-Arzelà Theorem 9.11.7 is useful in the theory of differential equations in proving Peano's Theorem on the existence of solutions of differential equations. We have already seen in Section 5.9 how Banach's Fixed Point Theorem 5.4 is used to establish Picard's Theorem 5.9.1, which guarantees that a differential equation

$$\frac{dy}{dx} = f(x,y)$$

with initial condition $y(x_0) = y_0$, where f is continuous on a closed rectangle $D \equiv \{(x,y) : |x-x_0| \leq a, |y-y_0| \leq b\}$ has a unique solution if f satisfies a Lipschitz condition with respect to y on D. Peano's Theorem guarantees the existence of solutions of such equations without requiring that f satisfy a Lipschitz condition with respect to y on D. Peano's Theorem, while being more general than Picard's, does not claim uniqueness. The proof of Picard's Theorem has the advantage that it indicates how the solution may be found by the iteration procedure given in Banach's Fixed Point Theorem 5.4. But the proof of Peano's Theorem only establishes existence and does not indicate in any way how a solution may be found.

In Section 5 we pointed to the significance of fixed point
theorems in the analysis of a variety of situations. The two outstanding
fixed point theorems are Brouwer's Fixed Point Theorem which states that
every continuous mapping of the closed unit ball of Euclidean n-space into
itself has a fixed point, and Schauder's Fixed Point Theorem which
generalises Brouwer's Theorem to infinite dimensional normed linear spaces.
The proof of Brouwer's Fixed Point Theorem uses algebraic topology and
Schauder's Fixed Point Theorem is developed from Brouwer's Theorem.
However, because our interest is in exhibiting an application of the
Ascoli-Arzelà Theorem 9.11.7 in proving Peano's Theorem, we will shorten
the classical proof by an appeal to Schauder's Theorem.

9.11.8 Schauder's Fixed Point Theorem. *Every continuous mapping of a compact convex subset of a normed linear space into itself has a fixed point.*

9.11.9 Peano's Theorem. *Consider the differential equation*

$$\frac{dy}{dx} = f(x,y)$$

with initial condition

$$y(x_0) = y_0$$

where f is continuous on a closed rectangle
$D \equiv \{(x,y) : |x-x_0| \leq a, \ |y-y_0| \leq b\}$ *in* $\mathbb{R}^2$. *Then on some closed interval*
$[x_0-r, x_0+r]$ *where* $0 < r < a$, *there exists a solution satisfying the initial condition.*

Proof. As in the proof of Picard's Theorem 5.9.1 we see that the differential equation with the initial condition is equivalent to the integral
equation

$$y(x) = y_0 + \int_{x_0}^{x} f(t,y(t))dt.$$

Since f is continuous on the bounded closed rectangle D, there exists a
k > 0 such that

$$|f(x,y)| \leq k \quad \text{for all } (x,y) \in D.$$

Choose $0 < r < \min\{a, \frac{b}{k}\}$.

Consider the set A of continuous functions $y(x)$ on $[x_0-r, x_0+r]$ such that

$$|y(x)-y_0| \leq kr \quad \text{for all } x \in [x_0-r, x_0+r] \tag{i}$$

and $\quad |y(x_1)-y(x_2)| \leq k|x_1-x_2| \quad$ for all $x_1, x_2 \in [x_0-r, x_0+r]$. $\quad$ (ii)

We show that A is a compact convex subset of the Banach space
$(C[x_0-r,x_0+r], \|\cdot\|_\infty)$:

Condition (i) tells us that A is a subset of the closed ball $B[y_0(x); kr]$
and so is bounded. Condition (ii) tells us that A is equicontinuous on
$[x_0-r, x_0+r]$. It is not difficult to prove that A is closed. Therefore
by the Ascoli-Arzelà Theorem 9.11.7, A is a compact subset of
$(C[x_0-r,x_0+r], \|\cdot\|_\infty)$. There is no difficulty in seeing that A is also
convex.

Consider the mapping T from A into $C[x_0-r,x_0+r]$ defined by

$$Ty(x) = y_0 + \int_{x_0}^{x} f(t, y(t)) \, dt.$$

We show that T maps A into itself:

For all $x \in [x_0-r, x_0+r]$,

$$|Ty(x)-y_0| = \left| \int_{x_0}^{x} f(t, y(t)) \, dt \right|$$

$$\leq k|x-x_0|$$

$$\leq kr,$$

and for all $x_1, x_2 \in [x_0-r, x_0+r]$,

$$|Ty(x_1)-Ty(x_2)| = \left| \int_{x_2}^{x_1} f(t, y(t)) \, dt \right|$$

$$\leq k|x_1-x_2|.$$

We next prove that T is a continuous mapping on A:

The rectangle D is compact in $(\mathbb{R}^2, \|\cdot\|_2)$ so by Heine's Theorem 9.3, f is
uniformly continuous on D which implies that given $\varepsilon > 0$ there exists a
$\delta > 0$ such that

$$\left| f(x,y) - f(x,y_1) \right| < \varepsilon$$

for all $(x,y), (x,y_1) \in D$ where $\left\| (x,y) - (x,y_1) \right\|_2 < \delta$. Given $y_1(x) \in A$ consider $y(x) \in A$ such that $\left\| y - y_1 \right\|_\infty < \delta$. Then $\left| y(x) - y_1(x) \right| < \delta$ for all $x \in [x_0-r, x_0+r]$. So

$$\left| Ty(x) - Ty_1(x) \right| = \left| \int_{x_0}^{x} \big(f(t,y(t)) - f(t,y_1(t)) \big) dt \right|$$

$$\leq \varepsilon \left| x - x_0 \right|$$

$$\leq \varepsilon r \quad \text{for all } x \in [x_0-r, x_0+r].$$

Therefore,

$$\left\| Ty - Ty_1 \right\|_\infty \leq \varepsilon r \quad \text{for } \left\| y - y_1 \right\|_\infty < \delta;$$

that is, T is continuous at y_1 which implies that T is continuous on A.

It now follows from Schauder's Theorem 9.11.8 that T has a fixed point in A which is a solution to the integral equation equivalent to the original differential equation. □

9.11.10 Remark. It is informative to notice that if given x_0, for every initial condition

$$y(x_0) = y_0$$

where $y_0 \in [c,d]$ there exists a unique solution to the differential equation, then this solution function varies continuously with $y_0 \in [c,d]$:

Consider the set P of solution functions $y(x)$ corresponding to the initial conditions $y(x_0) = y_0$ for all $y_0 \in [c,d]$ as a subset of the Banach space $(C[x_0-r, x_0+r], \|\cdot\|_\infty)$.

From the first step in the proof of Peano's Theorem 9.11.9 we see that given any $y_0 \in [c,d]$, the solution $y(x) \in B[y_0(x); kr]$. So for all $y(x) \in P$,

$$c - kr \leq y(x) \leq d + kr \quad \text{for all } x \in [x_0-r, x_0+r]$$

and therefore

$$\|y\|_\infty \leq \max\{|c-kr|,|d+kr|\} \quad \text{for all } y(x) \in P;$$

that is, P is bounded in $(C[x_0-r,x_0+r], \|\cdot\|_\infty)$.
But also, since every solution $y(x)$ satisfies

$$|y(x_1)-y(x_2)| \leq k|x_1-x_2| \quad \text{for all } x_1,x_2 \in [x_0-r,x_0+r]$$

we see that P is equicontinuous on $[x_0-r,x_0+r]$.
Furthermore P is closed. For if $\{y_n(x)\}$ is a sequence of solutions
converging uniformly to $y(x)$ on $[x_0-r,x_0+r]$ then from Theorem 7.35 since

$$\frac{dy_n}{dx} = f(x,y_n(x))$$

then $$\frac{dy}{dx} = f(x,y(x));$$

that is, $y(x)$ is also a member of P.
From the Ascoli-Arzelà Theorem 9.11.7 we deduce that P is compact.
Now since we have assumed uniqueness of solutions, given x_0, the mapping

$$y(x_0) \mapsto y_0$$

of P into the interval $[c,d]$ is one-to-one.

But $$|y_1(x_0)-y_2(x_0)| \leq \|y_1-y_2\|_\infty \quad \text{for any } y_1,y_2 \in P,$$

so this mapping is continuous. From Theorem 9.1 the inverse mapping is also
continuous and this establishes the continuous dependence of the solutions
on the initial conditions. $\square$

9.12 Exercises.

1.   (i)   Given a continuous real function f on (a,b) prove that f is uniformly continuous on (a,b) if and only if $\lim_{t \to a+} f(t)$ and $\lim_{t \to b-} f(t)$ both exist.

    (ii)   Given a differentiable real function f on $\mathbb{R}$ prove that f is uniformly continuous on $\mathbb{R}$ if there exists a $t_0 > 0$ and an $M > 0$ such that $|f'(t)| \leq M$ for all $|t| > t_0$.

        Give an example to show that the converse is not true in general.

    (iii)   Examine the following real functions for uniform continuity

      (a)   f on $(0,\infty)$ where

$$f(t) = \frac{\sin t}{t}$$

      (b)   f on $[1,\infty)$ where

$$f(t) = \ln t$$

      (c)   f on $[0,\infty)$ where

$$f(t) = \sqrt{t}.$$

2.   Given a sequence $\{f_n\}$ of monotone real functions defined on $[a,b]$ pointwise convergent to a continuous function f, prove that f is monotone and $\{f_n\}$ converges uniformly to f.

3.   (i)   In real $(C[a,b], \|\cdot\|_\infty)$ prove that if $f \in C[a,b]$ has the property that

$$\int_a^b f(t).t^n dt = 0 \quad \text{for all } n \in \mathbb{Z}^+$$

        then $f = 0$.

    (ii)   In real $(C[-\pi,\pi], \|\cdot\|_\infty)$ prove that if $f \in C[-\pi,\pi]$ has the property that

$$\int_{-\pi}^{\pi} f(t).\sin nt \, dt = 0$$

and

$$\int_{-\pi}^{\pi} f(t)\cos nt\, dt = 0 \quad \text{for all } n \in \mathbb{Z}^{+}$$

then $f = 0$.

4.   Consider the unit circle $\Gamma \equiv \{(\lambda,\mu) \in \mathbb{R}^2 : \lambda^2+\mu^2 = 1\}$.

(i)   Prove that the set $P_{\mathbb{Q}}(\Gamma)$ of polynomials on Γ with rational coefficients is a countable dense subset of real $(C(\Gamma), \|\cdot\|_\infty)$.

(ii)   Deduce, or otherwise prove, that real $(\tilde{C}(2\pi), \|\cdot\|_\infty)$ is separable.

5.   (i)   Prove that the linear span of the set of real functions $\{f_n : n \in \mathbb{N}\}$ where

$$f_n(t) = e^{nt} \quad \text{for } t \in [a,b]$$

is dense in real $(C[a,b], \|\cdot\|_\infty)$.

(ii)   Consider the closed unit disc $D \equiv \{z \in \mathbb{C} : |z| \leq 1\}$.
Prove that the linear span of the set of continuous complex functions f on D where

$$|f(z)| = 1 \quad \text{for all } z \in D$$

is dense in complex $(C(D), \|\cdot\|_\infty)$.

6.   For the real Banach space $(C(\mathbb{R}), \|\cdot\|_\infty)$ show that

(i)   Weierstrass Approximation Theorem does not hold and

(ii)   the space is not separable.

7. (i)   Show that the set of real functions $\{f_n : n \in \mathbb{N}\}$ on $[0,\pi]$ where

$$f_n(t) = \sin nt$$

is not equicontinuous on $[0,\pi]$.

   (ii)   Consider the sequence $\{f_n\}$ in $C(\mathbb{R}^+)$ defined by

$$f_n(t) = \sin\sqrt{t+4n^2\pi^2}.$$

Prove that $\{f_n\}$ is equicontinuous and pointwise convergent on $\mathbb{R}^+$ but that the closure of the set $\{f_n : n \in \mathbb{N}\}$ is not compact. (This example shows that the Ascoli-Arzelà Theorem does not necessarily hold in general for non-compact metric spaces $(X.d)$.)

8. (i)   Given a metric space (X,d), prove that if a subset A of $(C(X), \|\cdot\|_\infty)$ is equicontinuous on X then the closure $\overline{A}$ is equicontinuous on X.

   (ii)   Given a compact metric space (X,d), prove that a subset A of $(C(X), \|\cdot\|_\infty)$ is relatively compact if and only if A is bounded and equicontinuous;   (see Exercise 8.41.12).

9.   For a compact metric space (X,d), prove that a closed subset A of $(C(X), \|\cdot\|_\infty)$ is compact if A is equicontinuous and is pointwise bounded on X.

10. (i)   The sequence $\{f_n\}$ of differentiable real functions on $[0,1]$ have the property that, for some $M > 0$,

$$\left|f_n'(t)\right| \leqslant M \quad \text{for all } t \in [0,1] \text{ and all } n \in \mathbb{N}.$$

Prove that $\{f_n\}$ has a uniformly convergent subsequence.

   (ii)   Prove that a closed subset A of $(C^1[0,1], \|\cdot\|')$, (see Exercise 7.39.11(i)) is compact if and only if A is bounded and the set $\{f' : f \in A\}$ is equicontinuous on $[0,1]$.

11. (i)   In Exercise 7.39.8(i) we introduced the Fredholm operator K on
$C[a,b]$ defined by

$$(Kf)(x) = \int_a^b k(x,t) f(t) dt$$

where k is a continuous real function on the square
$\Box \equiv \{(x,t) : a \leqslant x \leqslant b, \ a \leqslant t \leqslant b\}$.
Prove that K is a compact operator on $(C[a,b], \|\cdot\|_\infty)$ by showing
that $K(B[0;1])$ is relatively compact.

(ii)   In Exercise 7.39.8(ii) we introduced the Volterra operator K
on $C[a,b]$ defined by

$$(Kf)(x) = \int_a^x k(x,t) f(t) dt$$

where k is a continuous real function on the triangle
$\Delta \equiv \{(x,t) : a \leqslant t \leqslant x, \ a \leqslant x \leqslant b\}$.
Prove that K is a compact operator on $(C[a,b], \|\cdot\|_\infty)$ by showing
that $K(B[0;1])$ is relatively compact.
(See Exercises 8.41.11(i) and 8.41.12(ii).)

V. THE METRIC TOPOLOGY

In our development of the analysis of metric spaces we have seen that concepts of limits of sequences, cluster points of sets, continuity of functions and compactness of sets are carried over from real and complex analysis through our notion of distance. Throughout this presentation we have used a sequential approach, building the definitions of our concepts on the basic idea of convergence of sequences.

However, as we developed our ideas we often formulated alternative characterisations which had little explicit reference to a sequential method and sometimes had no explicit reference to distance.

We now show that there is a further underlying structure in a metric space which is not explicitly framed in terms of distance and because of this we are able to extend our analysis to more general objects than metric spaces. This generalisation is another story, but in this chapter we identify what is called the topological structure of a metric space, look at its particular properties and provide alternative characterisations of our basic concepts in terms of this structure.

This is essentially an exercise of reviewing the material we have developed to show how it fits into the analysis of general topological spaces. Methods using the topological structure which we are to outline, are no more efficient in developing the theory we have presented and that is the reason why we have avoided them until now. The advantage of topological methods is that they lend themselves to generalisation. The achievements of the topological generalisation are better appreciated if they proceed from an acquaintance of the analysis of metric spaces treated from the more intuitive distance oriented approach.

10. THE TOPOLOGICAL ANALYSIS OF METRIC SPACES

The important sets for describing what is called the topological structure of a metric space are the open balls, (see Definition 2.1). We have from time to time shown how a concept can be characterised in terms of the open balls of the space, for example, when we discussed cluster points (Theorem 4.7), and ball cover compactness (Definition 8.31).

<u>10.1 Definition.</u> Given a metric space (X,d), a subset G is called an *open set* if for every $x \in G$ there exists an $r(x) > 0$ such that $B\big(x;r(x)\big) \subseteq G$.

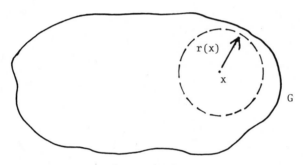

Figure 20. $B\big(x;r(x)\big) \subseteq G$.

The concept of an open set is derived from that of an open ball. It is important to explore more precisely the relation between open balls and open sets.

<u>10.2 Theorem.</u> *In a metric space* (X,d), *every open ball is an open set.*

<u>Proof.</u> Consider the open ball $B(x_0;r)$ and a point $x \in B(x_0;r)$. From Theorem 2.14 we see that there exists an $r'(x) > 0$ such that $B\big(x;r'(x)\big) \subseteq B(x_0;r)$. $\square$

This result also justifies our terminology. But further, open sets can be characterised in a different way by open balls.

<u>10.3 Theorem.</u> *Given a metric space* (X,d), *a subset* G *is open if and only if* G *is a union of open balls.*

Proof. Suppose that G is an open set. Then for each x ∈ G there exists an r(x) > 0 such that $B(x;r(x)) \subseteq G$. So

$$\cup\{B(x;r(x)) : x \in G\} \subseteq G \subseteq \cup\{B(x;r(x)) : x \in G\}.$$

Suppose that G = ∪{B : B ∈ F} where F is a family of open balls. Then for each x ∈ G there exists a B ∈ F such that x ∈ B. But from Theorem 10.2, B is an open set so there exists an r(x) > 0 such that

$$B(x;r(x)) \subseteq B \subseteq G$$

and therefore G is open. □

We now set out the fundamental properties of the family of open sets.

10.4 Theorem. *Given a metric space* (X,d),

   (i)   *both φ and X are open sets,*

  (ii)   *any union of open sets is open, and*

 (iii)   *any intersection of a finite number of open sets is open.*

Proof.   (i) Since there are no points in φ, each point in φ is the centre of an open ball contained in φ. For any x ∈ X, every open ball centred on X is contained in X.

      (ii) Consider a family $\{G_\alpha\}$ of open sets. By Theorem 10.3, each G_α is a union of open balls. Therefore $\cup G_\alpha$ is a union of open balls and so by Theorem 10.3 is an open set.

      (iii) Consider a finite family $\{G_k : k \in \{1,2,\ldots,n\}\}$ of open sets and $x \in \overset{n}{\underset{1}{\cap}} G_k$. Now $x \in G_k$ for each $k \in \{1,2,\ldots,n\}$. Since each G_k is open there exist $r_k(x) > 0$ such that $B(x;r_k(x)) \subseteq G_k$ for each $k \in \{1,2,\ldots,n\}$. Writing $r \equiv \min\{r_k(x) : k \in \{1,2,\ldots,n\}\}$ we have $B(x;r(x)) \subseteq G_k$ for each $k \in \{1,2,\ldots,n\}$; that is, $B(x;r(x)) \subseteq \overset{n}{\underset{1}{\cap}} G_k$. Therefore, $\overset{n}{\underset{1}{\cap}} G_k$ is open. □

10.5 Remark. The finiteness condition in Theorem 10.4(iii) is important because an intersection of an infinite family of open sets need not necessarily be open. In ℝ with the usual norm, consider the family of open intervals $\{(-\frac{1}{k},\frac{1}{k}) : k \in \mathbb{N}\}$.   Now $\cap\{(-\frac{1}{k},\frac{1}{k}) : k \in \mathbb{N}\} = \{0\}$, a single point set which is not an open set. $\square$

10.6 Example.   In $\mathbb{R}$ with the usual norm,

   (i)   an open interval (a,b) is an open ball $B(\frac{a+b}{2};\frac{b-a}{2})$ so is an open set,

   (ii)   a union of open intervals is an open set, (Theorem 10.4(ii)),

   (iii)   a non-open interval, say (a,b] is not an open set because, for every r > 0

$$B(b;r) = (b-r,b+r) \not\subseteq (a,b],$$

   (iv)   single point sets are not open. $\square$

The discrete metric space provides a check on our intuition.

10.7 Example.   For any non-empty set X with the discrete metric,

   (i)   every single point set is an open set because, for any $x \in X$, $\{x\} = B(x;1)$ an open ball and so an open set,

   (ii)   all subsets are open sets, (Theorem 10.4(ii)). $\square$

10.8 Example.   In $(C[a,b], \|\cdot\|_\infty)$, the subset A consisting of functions f such that

$$\left|\int_a^b f(t)dt\right| < 1$$

is open:
Consider any $f_0 \in A$, then $\left|\int_a^b f_0(t)dt\right| \equiv k < 1$.

Choose $r(f_0) \equiv \frac{1-k}{b-a}$.   Then for any $f \in B(f_0;r(f_0))$ we have

$$\|f-f_0\|_\infty < \frac{1-k}{b-a}$$

and so

$$\left|\int_a^b f(t)dt\right| \leq \left|\int_a^b \left(f(t)-f_0(t)\right)dt\right| + \left|\int_a^b f_0(t)dt\right|$$

$$< \frac{1-k}{b-a} \cdot (b-a) + k = 1.$$

Therefore $B\left(f_0;r(f_0)\right) \subseteq A$ and we conclude that A is open. $\square$

10.9 Definition. Given a metric space (X,d), the family of open sets is called the *metric topology* and is denoted by T_d.
In a normed linear space $(X, \|\cdot\|)$ the metric topology generated by the norm is called the *norm topology* and is denoted by $T_{\|\cdot\|}$.
It is generally difficult to visualise the family of open sets in a metric space. This contrasts with the simplicity of having one function defining distance.

In $\mathbb{R}$ with the usual norm, the norm topology does have some clarifying specification.

10.10 Theorem. *In $\mathbb{R}$ with the usual norm, every open set is the union of a countable family of disjoint open intervals.*

Proof. Consider an open subset G. For each $x \in G$ there exists an $r(x) > 0$ such that $(x-r,x+r) \subseteq G$. Denote by I_x the union of all open intervals containing x and contained in G. If $y \in I_x$ and $x \neq y$ then $I_x = I_y$. If $x,y \in G$, $x \neq y$ and $z \in I_x \cap I_y$ then $I_z = I_x$ and $I_z = I_y$ so $I_x = I_y$. So we have proved that each point $x \in G$ belongs to a uniquely determined open interval and such open intervals are disjoint.
Now consider the set of rationals $\mathbb{Q}$ in sequence. To each open interval assign the first member of this sequence belonging to that interval. Since the family of open intervals is disjoint this mapping is one-to-one. So the family of open intervals is countable. $\square$

We are helped to understand the metric topology by recalling from Theorem 10.3 that every open set is a union of open balls. So the family of open balls is a subfamily of the metric topology which generates that topology by unions. To be able to specify a manageable subfamily which generates the metric topology in this way is a considerable advantage.

10.11 Definition. Given a metric space (X,d), a subfamily B of the metric topology T_d is called a *base* for T_d if every open set is a union of sets from B.

A useful characterisation of a base is given in the following theorem.

10.12 Theorem. *Given a metric space (X,d), a subfamily of open sets B is a base for the metric topology T_d if and only if for every open set G and every $x \in G$ there exists a set $B_x \in B$ such that $x \in B_x \subseteq G$.*

Proof. Suppose that B is a base for T_d; that is, every open set G is a union of sets in B. Then for every $x \in G$ there exists a set $B_x \in B$ such that $x \in B_x \subseteq G$.

Suppose that for every open set G and every $x \in G$ there exists a $B_x \in B$ such that $x \in B_x \subseteq G$. Then

$$\cup\{B_x : x \in G\} \subseteq G \subseteq \cup\{B_x : x \in G\},$$

so G is a union of sets in B; that is, B is a base for T_d. $\square$

10.13 Remark. Using this terminology we say that for any metric space, the family of open balls is a base for the metric topology. $\square$

The homogeneity of a normed linear space shown in Theorem 2.4 enables us to give a particularly simple description of the way the norm topology is generated.

10.14 Theorem. *Given a normed linear space $(X, \|\cdot\|)$, a base for the norm topology $T_{\|\cdot\|}$ is given by translates of strictly positive multiples of the open unit ball.*

The notion of a base for a metric topology enables us to give a topological characterisation of the notion of equivalent metrics.

10.15 Lemma. *Given a non-empty set X, metrics d and d' for X generate the same metric topology if and only if for every $x \in X$ and $r > 0$ there exists an $s(x) > 0$ and an $s'(x) > 0$ such that*

$$B_{d'}\big(x;s'(x)\big) \subseteq B_d(x;r) \qquad\qquad\qquad\qquad \text{(i)}$$

and $\qquad B_d\big(x;s(x)\big) \quad \subseteq B_{d'}(x;r) \qquad\qquad\qquad\qquad \text{(ii)}$

<u>Proof</u>. Suppose that $T_d = T_{d'}$. Then the open balls in (X,d) are open sets in (X,d') and the open balls in (X,d') are open sets in (X,d). Now the open balls are a base for the metric topology so for $B_d(x;r)$ we have from Theorem 10.12 that there exists an $s'(x) > 0$ such that

$$B_{d'}\big(x;s'(x)\big) \subseteq B_d(x;r)$$

and similarly for $B_{d'}(x;r)$ there exists an $s(x) > 0$ such that

$$B_d\big(x;s(x)\big) \subseteq B_{d'}(x;r).$$

Conversely, suppose that conditions (i) and (ii) hold. Consider an open set G in (X,d). Since G is open, for every $x \in G$ there exists an $r(x) > 0$ such that

$$B_d\big(x;r(x)\big) \subseteq G.$$

But from (i) there exists an $s'(x) > 0$ such that

$$B_{d'}\big(x;s'(x)\big) \subseteq B_d\big(x;r(x)\big) \subseteq G.$$

Then G is open in (X,d') which implies that $T_d \subseteq T_{d'}$. The symmetry of the argument, using (ii) proves that $T_{d'} \subseteq T_d$ and we conclude that $T_d = T_{d'}$. $\square$

Because it is the open unit ball which generates the norm topology, Lemma 10.15 takes a simple form for normed linear spaces.

<u>10.16 Corollary</u>. *Given a linear space* X, *norms* $\|\cdot\|$ *and* $\|\cdot\|'$ *for* X *generate the same norm topology if and only if there exist* $s,s' > 0$ *such that*

$$s'B_{\|\cdot\|'}(\underline{0};1) \subseteq B_{\|\cdot\|}(\underline{0};1) \qquad\qquad\qquad\qquad \text{(i)}$$

and $\qquad s\, B_{\|\cdot\|}(\underline{0};1) \quad \subseteq B_{\|\cdot\|'}(\underline{0};1) \qquad\qquad\qquad\qquad \text{(ii)}$

<u>10.17 Theorem</u>. *Given a non-empty set* X, *metrics* d *and* d' *for* X *are equivalent if and only if* d *and* d' *generate the same metric topology.*

<u>Proof</u>. Suppose that $T_d = T_{d'}$. Then by Lemma 10.15 for any x ϵ X, given $\epsilon > 0$ there exists an s > 0 such that

$$B_d(x;s) \subseteq B_{d'}(x;\epsilon).$$

Consider a sequence $\{x_n\}$ convergent to x in (X,d). Then there exists a $\nu \in \mathbb{N}$ such that

$$x_n \in B_d(x;s) \qquad \text{for all } n > \nu.$$

Therefore,

$$x_n \in B_{d'}(x;\epsilon) \qquad \text{for all } n > \nu;$$

that is, $\{x_n\}$ is convergent to x in (X,d'). Symmetry of the argument completes the proof that d and d' are equivalent metrics.

Suppose that d and d' do not generate the same metric topology on X. Then by Lemma 10.15, for at least one x ϵ X and one of the metrics d, there exists an r > 0 such that

$$B_{d'}(x;\tfrac{1}{n}) \setminus B_d(x;r) \neq \phi \qquad \text{for every } n \in \mathbb{N}.$$

For each n $\in \mathbb{N}$, choose $x_n \in B_{d'}(x;\tfrac{1}{n}) \setminus B_d(x;r)$. Then sequence $\{x_n\}$ converges to x in (X,d') but since $x_n \in C\big(B_d(x;r)\big)$ for all n $\in \mathbb{N}$, $\{x_n\}$ does not converge to x in (X,d); that is, d and d' are not equivalent metrics. $\square$

<u>10.18 Remark</u>. In Corollary 3.20 we noted that on any given finite dimensional linear space all norms are equivalent. Topologically this corollary says that any given finite dimensional linear space has a unique norm topology. $\square$

We now see that there is a close link between the open sets and the closed sets in a metric space.

10.19 Theorem. *Given a metric space* (X,d)*, a subset* A *is open if and only if its complement* C(A) *is closed.*

Proof. In view of Theorems 4.18(i) and 10.4(i) it is sufficient to consider the case when A is a proper subset of X.

Suppose that A is open. Then for any $x \in$ A there exists an $r > 0$ such that $B(x;r) \subseteq$ A. But then $B(x;r) \cap C(A) = \phi$ and so from Theorem 4.15, C(A) is closed.

Suppose that C(A) is closed. Then again from Theorem 4.15 for any $x \in$ A there exists an $r > 0$ such that $B(x;r) \cap C(A) = \phi$. But then $B(x;r) \subseteq$ A; that is, A is open. $\square$

This theorem with Theorem 10.10 gives us some insight into the nature of closed sets in $\mathbb{R}$ with the usual norm.

10.20 Corollary. *In* $\mathbb{R}$ *with the usual norm, every closed set is the complement of the union of a countable family of disjoint open intervals.*

Theorem 10.19 is important in determining the openness or closedness of a set.

10.21 Example. Cantor's ternary set T is defined as that subset of $[0,1]$ with the usual metric consisting of those numbers which have a ternary representation with digits 0 and 2; (see Exercise 4.51.6). Geometrically, we can see that T is formed by removing a countable family of disjoint open intervals from $\mathbb{R}$:

$$G_0^1 \equiv (-\infty,0), \quad G_0^2 \equiv (1,\infty)$$
$$G_1^1 \equiv (\tfrac{1}{3},\tfrac{2}{3})$$

(those which must have 1 in the first ternary place),

$$G_2^1 \equiv (\tfrac{1}{9},\tfrac{2}{9}), \quad G_2^2 \equiv (\tfrac{7}{9},\tfrac{8}{9})$$

(those which must have 1 in the second ternary place),

$$G_3^1 \equiv (\tfrac{1}{27},\tfrac{2}{27}), \ G_3^2 \equiv (\tfrac{7}{27},\tfrac{8}{27}), \ G_3^3 \equiv (\tfrac{22}{27},\tfrac{23}{27}), \ G_3^4 \equiv (\tfrac{25}{27},\tfrac{26}{27})$$

(those which must have 1 in the third ternary place),

. . .

and so we see quite obviously that T is closed. □

10.22 Remark. In Theorems 4.18 and 10.4 we set out the fundamental
properties of the family of closed sets and the family of open sets. It
is quite clear that the one can now be derived from the other using
Theorem 10.19 and de Morgan's Theorem. □

We notice in Theorems 4.18(i) and 10.4(i) that in any metric
space (X,d), ϕ and X are both open and closed. In a discrete metric space
it follows from Example 10.7 and Theorem 10.19 that all subsets are both
open and closed. However, in normed linear spaces the situation fits more
with our intuition.

10.23 Theorem. *In a normed linear space* $(X, \|\cdot\|)$, *the only subsets which
are both open and closed are* ϕ *and* X.

Proof. Suppose that there exists a proper non-empty subset A of X which is
both open and closed. Then by Theorem 10.19, $C(A)$ is also both open and
closed. Consider $x \in A$ and $y \in C(A)$ and the set $\{\lambda x + (1-\lambda)y : 0 \le \lambda \le 1\}$.
Now the set of real numbers $\{\lambda : \lambda x + (1-\lambda)y \in C(A)\}$ is non-empty and is
bounded above by 1, so by the Supremum Axiom for $\mathbb{R}$ there exists a
$\lambda_0 \equiv \sup\{\lambda : \lambda x + (1-\lambda)y \in C(A)\}$ and $0 \le \lambda_0 \le 1$. Since λ_0 is an upper
bound for $\{\lambda : \lambda x + (1-\lambda)y \in C(A)\}$ then either $\lambda_0 x + (1-\lambda_0)y \in A$ or is a
cluster point of $\{\lambda x + (1-\lambda)y : 0 \le \lambda \le 1\} \cap A$ which implies that it is a
cluster point of A. Also since λ_0 is the least upper bound for
$\{\lambda : \lambda x + (1-\lambda)y \in C(A)\}$ then either $\lambda_0 x + (1-\lambda_0)y \in C(A)$ or is a cluster
point of $C(A)$. But both A and $C(A)$ are closed so $\lambda_0 x + (1-\lambda_0)y \in A \cap C(A)$
which is a contradiction. □

We noticed in Section 4 that in any metric space it is useful
to associate with any subset a related closed subset called its closure.
In view of Theorem 10.19 it is natural that we also associate with any
subset a related open set.

10.24 Definition. Given a metric space (X,d), for any subset A of X, a point $x \in X$ is called an *interior point* of A if there exists an $r(x) > 0$ such that

$$B\big(x;r(x)\big) \subseteq A.$$

For any point $x \in X$, a subset A of X is called a *neighbourhood* of x if x is an interior point of A.

10.25 Remark. Using this terminology with Definition 10.1 we can say that a subset A of a metric space (X,d) is open if and only if every point of A is an interior point of A, or A is open if and only if A is a neighbourhood of each of its points. □

10.26 Definition. Given a metric space (X,d), the *interior* of a subset A is the set of interior points of A and is denoted by int A.

The general properties for the interior of a set are complementary to those for closure given in Theorem 4.28.

10.27 Theorem. *Given a subset A of a metric space* (X,d),

(i)   int A *is open*,

(ii)   int A *is the largest open set contained in* A.

Proof.   (i) Consider $x \in$ int A. There exists an $r(x) > 0$ such that $B\big(x;r(x)\big) \subseteq A$. But since $B\big(x;r(x)\big)$ is an open set, for every $y \in B\big(x;r(x)\big)$ there exists an $r(y) > 0$ such that $B\big(y;r(y)\big) \subseteq B\big(x;r(x)\big) \subseteq A$. So y is an interior point of A. Therefore, $B\big(x;r(x)\big) \subseteq$ int A;   that is, int A is open.

(ii) Consider any open set G such that $G \subseteq A$. For any $x \in G$ there exists an $r(x) > 0$ such that $B\big(x;r(x)\big) \subseteq G \subseteq A$, so x is an interior point of A. Therefore $G \subseteq$ int A. □

10.28 Remark. In terms of the notion of interior we can say that a subset A of a metric space (X,d) is open if and only if int $A = A$. From Theorem 10.17 and Theorem 10.27(i) it is obvious that the interior of a set is invariant under equivalent metrics. □

It is important to see that the interior of a set is determined both by the underlying set of points in the metric space and the particular metric.

10.29 Example. For the closed interval $A \equiv [0,1]$,

    (i)   considered as a subset of $\mathbb{R}$ with the usual metric, int $A = (0,1)$,

    (ii)   considered as a subset of $\mathbb{R}$ with the discrete metric, int $A = [0,1]$,

   (iii)   considered as a subset of $(\mathbb{R}^2, \|\cdot\|_2)$, int $A = \phi$. $\square$

However, for normed linear spaces we have the following general property concerning the interior of subspaces.

10.30 Theorem. *For any proper linear subspace* M *of a normed linear space* $(X, \|\cdot\|)$, *we have* int $M = \phi$.

Proof. Since M is a proper subset there exists an $x_0 \in X \setminus M$ and since M is a linear subspace $\lambda x_0 \in X \setminus M$ for all real $\lambda \neq 0$. But given $\varepsilon > 0$, there exists a $\lambda' > 0$ such that $\lambda' x_0 \in B(\underline{0}; \varepsilon)$ so $B(\underline{0}; \varepsilon) \not\subseteq M$ and 0 is not an interior point of M. Given any point $x \in M$ we have that $x + \lambda' x_0 \in B(x; \varepsilon)$. But since M is a linear subspace $x + \lambda' x_0 \notin M$ so $B(x; \varepsilon) \not\subseteq M$ and x is not an interior point of M. Therefore, int $M = \phi$.

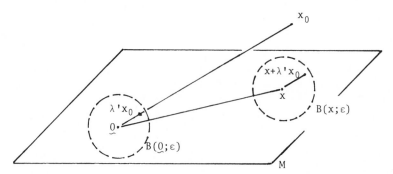

Figure 21. $\lambda' x_0 \in B(\underline{0}; \varepsilon) \setminus M$ and $x + \lambda' x_0 \in B(x; \varepsilon) \setminus M$ for all $x \in M$. $\square$

This result has wider application in normed linear spaces.

10.31 Corollary. *In a normed linear space* $(X, \|\cdot\|)$, *any subset* A *such that for* $x_0 \in A$, $sp(A-x_0)$ *is a proper linear subspace of* X, *has int* A = ϕ.

Proof. By Theorem 10.30 we have that int $sp(A-x_0) = \phi$. But from the homogeneity of the norm topology int A = x_0 + int $sp(A-x_0) = \phi$.

10.32 Example. In the normed linear space $(C[0,1], \|\cdot\|_1)$, the subset $A \equiv \{f \in C[0,1] : f(0) = 1, \|f\|_1 < 1\}$ has int A = ϕ:
The coset $\{f \in C[0,1] : f(0) = 1\}$ is the translate of a proper linear subspace which by Corollary 10.31 has empty interior and so A as a subset of this coset has empty interior. □

Because of the complementary relation between open and closed sets we would expect that the characterisation of a continuous mapping between metric spaces given in Theorem 6.19 in terms of closed sets would have a similar statement in terms of open sets.

10.33 Theorem. *Given metric spaces* (X,d) *and* (Y,d') *a mapping* $T : X \to Y$ *is continuous on* X *if and only if for every open set* G *in* (Y,d') *the set* $T^{-1}(G)$ *is open in* (X,d); (that is, a mapping is continuous if and only if the inverse images of open sets are open).

Proof. The proof follows directly from Theorem 6.19 using Theorem 10.19 and noticing that for any subset E in (Y,d'),

$$T^{-1}(C(E)) = C(T^{-1}(E)).$$

The proof could of course be developed directly from the ε-δ definition of continuity, Definition 6.1. □

We have a similar characterisation theorem for homeomorphisms as that given in Theorem 6.29.

10.34 Corollary. *A one-to-one mapping* T *from a metric space* (X,d) *onto a metric space* (Y,d') *is a homeomorphism if and only if for every open set* E

in (X,d) *the set* T(E) *is open in* (Y,d') *and for every open set* G *in*
(Y,d') *the set* T^{-1}(G) *is open in* (X,d); (that is, a homeomorphism onto is
a one-to-one mapping which preserves open sets).

<u>10.35 Remark</u>. It is Theorem 10.33 characterising continuity in terms of
the metric topology which points towards the possibility of the analysis
of a more general structure than a metric space and one which does not
depend on the notion of distance. A general topological space is a non-
empty set with a family of subsets which we specify as open sets and
which obey the fundamental properties established for the metric topology
in Theorem 10.4. Between such spaces we can define a continuous mapping
as one which obeys the property given in Theorem 10.33. ☐

In Definition 8.31 we introduced the idea of ball cover
compactness. This is the form of compactness which we now generalise to a
completely topological characterisation.

<u>10.36 Definition</u>. Given a metric space (X,d), a subset A is said to be
topologically compact if every cover of A by open sets has a finite
subcover.

We now show that this form of compactness is equivalent to the
other forms introduced in Section 8.

<u>10.37 Theorem</u>. *Given a metric space* (X,d), *a subset* A *is compact if and
only if* A *is topologically compact.*

<u>Proof</u>. From Theorem 10.2, open balls are open sets so if A is topologically
compact then A is ball cover compact.

Conversely, suppose that A is compact and consider a cover of A
by a family of open sets $\{G_\alpha\}$. Now from Theorem 10.3 each G_α is a union of
open balls so the family $\{G_\alpha\}$ provides a cover for A by open balls with
centres in A. But A is ball cover compact so there exists a finite
subfamily of the open balls making up $\cup G_\alpha$ which covers A. To each such
open ball associate a set from the open cover containing it. Then we see
that there is a finite subcover from $\{G_\alpha\}$; that is, A is topologically
compact. ☐

There is a complementary form of topological compactness stated in terms of closed sets. For simplicity we confine ourselves to compact metric spaces; it is in such a setting that this characterisation is applied.

10.38 Theorem. *A metric space* (X,d) *is compact if and only if every family of closed sets where every finite subfamily has non-empty intersection has itself non-empty intersection.*

Proof. For any family $\{F_\alpha\}$ of closed sets with empty intersection the family $\{C(F_\alpha)\}$ is a cover for X by open sets. Now X is compact if and only if $\{C(F_\alpha)\}$ has a finite subcover; that is, there exists a subfamily $\{F_k : k \in \{1,2,\ldots,n\}\}$ such that $X \subseteq \cup\{C(F_k) : k \in \{1,2,\ldots,n\}\}$. From de Morgan's Theorem this is equivalent to $\cap\{F_k : k \in \{1,2,\ldots,n\}\} = \phi$. $\Box$

10.39 Remark. It is clear that compactness in the form of topological compactness can be defined for more general topological spaces. If in topological spaces we use Theorem 10.19 to define a closed set as the complement of an open set then the form of compactness given in Theorem 10.38 can be seen to be equivalent to topological compactness in topological spaces. $\Box$

In our presentation of the analysis of metric spaces we have relied heavily on sequential methods. There is some difficulty in extending such methods to more general structures. The appropriateness of sequential methods in metric spaces is due to the particular property that the metric topology has a countable local base.

10.40 Theorem. *Given a metric space* (X,d) *and a point* $x \in X$ *for any open neighbourhood* G *of* x *there exists an* $n \in \mathbb{N}$ *such that* $x \in B(x;\frac{1}{n}) \subseteq G$.

10.41 Corollary. *Given a metric space* (X,d), *a base for the metric topology is the family of all open balls with radius* $\frac{1}{n}$ *for all* $n \in \mathbb{N}$.

It should be recognised that separability can be characterised by a stronger countability property of the metric topology.

228.   The metric topology

10.42 Theorem. *A metric space* (X,d) *is separable if and only if the metric topology* T_d *has a countable base.*

Proof. Suppose that T_d has a countable base $\{B_n : n \in \mathbb{N}\}$. For each $n \in \mathbb{N}$, choose $x_n \in B_n$. We show that $\{x_n : n \in \mathbb{N}\}$ is dense in (X,d):

For any $x \in X$, given $\varepsilon > 0$ there exists an $n_0 \in \mathbb{N}$ such that $x \in B_{n_0} \subseteq B(x;\varepsilon)$. But $x_{n_0} \in B_{n_0}$ so $B(x;\varepsilon) \cap \{x_n : n \in \mathbb{N}\} \neq \phi$ which implies that $\{x_n : n \in \mathbb{N}\}$ is dense in (X,d).

Conversely, suppose that (X,d) is separable and that the countable set $\{x_n : n \in \mathbb{N}\}$ is dense in (X,d). Consider the family of open balls $\{B(x_n;q) : n \in \mathbb{N}, q \in \mathbb{Q}\}$. This family is countable; we show that it is a base for the metric topology T_d:

For any $x \in X$ and any open set G containing x, there exists an $r > 0$ such that $B(x;r) \subseteq G$. Consider $B(x;\frac{r}{3})$. Now since $\{x_n : n \in \mathbb{N}\}$ is dense in (X,d), there exists an $x_{n_0} \in B(x;\frac{r}{3})$. Since the rationals $\mathbb{Q}$ are dense in the reals $\mathbb{R}$, there exists a $q \in \mathbb{Q}$ such that $\frac{r}{3} < q < \frac{2r}{3}$, so

$$x \in B(x_{n_0};q) \subseteq B(x;r)$$

Then by Theorem 10.12, we have that $\{B(x_n;q) : n \in \mathbb{N}, q \in \mathbb{Q}\}$ is a base for T_d.

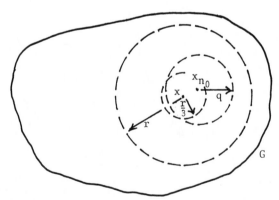

Figure 22. $x \in B(x_{n_0};q) \subseteq B(x;r) \subseteq G.$ $\square$

A rich topology is one which is abundant in open sets and one which satisfies what are called separation properties. The success of the sequential method in metric spaces depends fundamentally on the property proved in Theorem 3.4(i) concerning the uniqueness of limits of convergent sequences. This property is guaranteed for metric spaces because there are sufficient open sets to separate points; in particular, for any metric space (X,d) give any $x,y \in X$, $x \neq y$, there exists an $r > 0$ such that $y \notin B(x;r)$. In terms of the topology this says that for any $x,y \in X$, $x \neq y$, there exists an open neighbourhood G of x such that $y \notin G$. It is this topological property which assures us that in any metric space, single point sets are closed sets.

This basic separation property is strengthened by the following result.

10.43 Theorem. *Given a metric space* (X,d), *for any disjoint closed subsets* A *and* B *there exist disjoint open subsets* G *and* H *such that* $A \subseteq G$ *and* $B \subseteq H$.

Proof. If either A or B is empty then ϕ and X are open disjoint subsets which suffice. Suppose both A and B are non-empty. Consider $a \in A$. Since $a \notin B$ and B is closed we have from Exercise 4.51.12(i) that $d(a,B) \equiv r(a) > 0$. Similarly for $b \in B$, $d(b,A) \equiv r(b) > 0$.

We define $G \equiv \cup\left\{B\left(a;\dfrac{r(a)}{3}\right) : a \in A\right\}$

and $H \equiv \cup\left\{B\left(b;\dfrac{r(b)}{3}\right) : b \in B\right\}$.

Now both G and H as the union of open balls are open sets. Also $A \subseteq G$ and $B \subseteq H$. We need to show that G and H are disjoint:
Suppose $x \in G \cap H$ then there exists an $a_0 \in A$ and a $b_0 \in B$ such that
$$x \in B\left(a_0;\frac{r(a_0)}{3}\right) \cap B\left(b_0;\frac{r(b_0)}{3}\right).$$

So $d(a_0,b_0) \leqslant d(a_0,x) + d(b_0,x)$

$$< \frac{r(a_0)}{3} + \frac{r(b_0)}{3}$$

$$\leqslant \tfrac{2}{3}d(A,B)$$

which is impossible. $\square$

The topological name for the property established for metric spaces in Theorem 10.43 is called the *normal* separation property. Being normal also guarantees that for any metric space (X,d), the real linear space $C(X)$ is always substantial enough to separate disjoint closed subsets of X by continuous functions. We establish this property directly.

10.44 Theorem. *Given a metric space* (X,d), *and any disjoint closed subsets* A *and* B *there exists a continuous function* $f : X \to [0,1]$ *such that*

$$
\begin{aligned}
f(x) &= 0 \quad \textit{for all } x \in A \\
&= 1 \quad \textit{for all } x \in B
\end{aligned}
$$

Proof. Since A and B are closed and disjoint,

$$d(x,A) + d(x,B) > 0 \quad \text{for all } x \in X.$$

Define f on X by

$$f(x) = \frac{d(x,A)}{d(x,A)+d(x,B)} \cdot$$

Since the mapping $x \mapsto d(x;A)$ is continuous on X, (see Exercise 6.39.9(i)), f is continuous on X and f satisfies the condition of the theorem. □

10.45 Remark. The topological property in Theorem 10.44 is called the *Urysohn property*. Theorem 10.44 actually provides a proof of Theorem 10.43 using the topological characterisation of continuity, Theorem 10.33. □

10.46 Exercises.

1.   In the spaces given determine whether the following subsets
are open.

   (i)   In $\mathbb{R}$ with the usual norm

      (a)   the union of open intervals $(\frac{1}{n+1}, \frac{1}{n})$ for all $n \in \mathbb{N}$,

      (b) $\{0 < x < 1 : x$ has a decimal representation with 0 as
                         first digit$\}$,

      (c) $\{\frac{1}{x} : x \neq 0\}$.

   (ii)   In $(\mathbb{R}^3, \|\cdot\|_2)$,

      (a) $\{x \in \mathbb{R}^3 : \|x\|_\infty < 1\}$,

      (b) $\{(\lambda, \mu, \nu) \in \mathbb{R}^3 : -1 < \lambda + \mu + \nu < 1\}$,

      (c) $\{x \in \mathbb{R}^3 : \|x\| > 1$ for any norm $\|\cdot\|$ on $\mathbb{R}^3\}$.

   (iii)   In $(C[0,1], \|\cdot\|_1)$,

      (a)   the set of all continuous real functions f where
            $\sup\{|f(t)| : t \in [0,1]\} < 1$,

      (b)   the set of all positive real functions f where
            $\|f\|_1 < 1$,

      (c)   the set of all strictly positive real functions f.

2. (i)   In $\mathbb{R}$ with the Zero Bias metric d_0, (see Exercise 1.34.5), show
           that $(-1,0]$ is an open set.

   (ii)   In $\mathbb{R}^2$ with the Post Office metric d_p, (see Exercise 1.34.6),
           prove that all single point sets except $(0,0)$ are open.

   (iii)   In a metric space (X,d) where the metric d satisfies the ultra-
           metric inequality (see Exercise 2.33.5), prove that

      (a)   any open ball is both an open and closed set, and
      (b)   any closed ball is both an open and closed set.

3. (i)   Prove that, in any metric space

   (a)   the complement of every single point set is open,

   (b)   the complement of every finite set is open.

   (ii)   (a)   For any metric space, prove that every subset is open if
              and only if every single point set is open.

           (b)   Prove that in any non-trivial normed linear space there
              exist subsets which are not open.

4.   Given a metric space (X,d) and a subset Y of X, prove that

   (i)   a subset A of Y is open in $(Y,d|_Y)$ if and only if there exists
           an open set G in X such that $A = G \cap Y$,

   (ii)   every open subset of $(Y,d|_Y)$ is open in (X,d) if and only if Y
           is open in (X,d).

5.   In the spaces given determine the interiors of the following
subsets.

   (i)   In $(\mathbb{R}^2, \|\cdot\|_2)$,

           (a) $\{(\lambda,\mu) : 0 < \lambda < 1,\ \mu \in \mathbb{Q}\}$,

           (b) $\{(\lambda,\mu) : 0 < \lambda+\mu \leqslant 1\}$,

           (c) $\{x \in \mathbb{R}^2 : \|x\| \leqslant 1$ for any norm $\|\cdot\|$ on $\mathbb{R}^2\}$.

   (ii)   In complex $(c, \|\cdot\|_\infty)$,

           (a) c_0 the set of sequences which converge to 0,

           (b)   the set of sequences which converge to real limits,

           (c)   the set of sequences which converge to limits in the
              closed unit disc $\{\lambda \in \mathbb{C} : |\lambda| \leqslant 1\}$.

   (iii)   In real $(\mathcal{B}(\mathbb{R}), \|\cdot\|_\infty)$,

           (a)   the set of functions such that
              $f(t) \to 0$ as $|t| \to \infty$,

           (b)   the set of functions such that
              $0 \leqslant f(t) \leqslant 1$ for all $t \in \mathbb{R}$,

           (c)   the set of odd functions; that is, where
              $f(-t) = -f(t)$ for all $t \in \mathbb{R}$.

6. (i)   Prove that in any normed linear space

           int $S(\underset{\sim}{0};1) = \phi$.

  (ii)   Give an example to show that in a metric space (X,d) it is not
           necessarily true that spheres have empty interior.

7. (i)   For a normed linear space $(X, \|\cdot\|)$, prove that the interior of a
           convex set is always convex.

  (ii)   For a linear space X with a metric d, determine whether the
           interior of a convex set is always convex.

8.   For any subset A of a metric space (X,d), prove that

    (i)   int $C(A) = C(\overline{A})$,

   (ii)   int $A \cap \partial A = \phi$,

  (iii) $\overline{A} = (\text{int } A) \cup \partial A$.

   (iv) A is both open and closed if and only if $\partial A = \phi$.

9.   For subsets A and B of a metric space (X,d) prove that

    (i)   if $A \subseteq B$ then int $A \subseteq$ int B.

   (ii) $\text{int}(A \cap B) = \text{int } A \cap \text{int } B$.

  (iii) $\text{int}(A \cup B) \supseteq \text{int } A \cup \text{int } B$, but give an example to show that
           equality does not hold in general.

10. (i)   Prove that any linear functional f on a normed linear space
           $(X, \|\cdot\|)$ maps open sets to open sets.

  (ii)   Given normed linear spaces $(X, \|\cdot\|)$ and $(Y, \|\cdot\|')$ where Y is
           finite dimensional, and a linear mapping T from X onto Y.

           (a)   Prove that T maps open sets to open sets.
           (b)   Prove that if ker T is closed then T is continuous.

11.   Consider metric spaces (X,d) and (Y,d') and the product space $(X \times Y, d_\pi)$, (see Exercise 1.34.11(i)).

(i)   Prove that if B is a base for the metric topology T_d and B' is a base for the metric topology $T_{d'}$, then $\{B \times B' : B \in B, \ B' \in B'\}$ is a base for the product topology T_π.

(ii)   Prove that the projection mappings
$p_1 : X \times Y \to X$ where $p_1(x,y) = x$ and
$p_2 : X \times Y \to Y$ where $p_2(x,y) = y$

map open sets to open sets but show that they do not in general map closed sets to closed sets.

12.   Given a normed linear space $(X, \|\cdot\|)$ and a closed linear subspace M, consider the quotient space $\left(\frac{X}{M}, \|\cdot\|'\right)$ and the quotient mapping $\pi : X \to \frac{X}{M}$ defined by

$$\pi(x) = x + M,$$

(see Exercises 4.51.21 and 7.39.15).

(i)   Prove that the quotient mapping π is linear, continuous and maps open sets to open sets.

(ii)   Prove that if B is a base for the norm topology $T_{\|\cdot\|}$ then $\{\pi(B) : B \in B\}$ is a base for the quotient topology.

13. (i)   Prove that in a metric space (X,d), for any closed subset F and open subset E such that $F \subseteq E$, there exists an open subset G such that

$$F \subseteq G \subseteq \overline{G} \subseteq E.$$

(ii)   Deduce that if B is a base for the metric topology T_d for a metric space (X,d) then, for any $B \in B$ and $x \in B$ there exists a $B_1 \in B$ such that

$$x \in \overline{B_1} \subseteq B.$$

APPENDICES

This section contains an outline of background material in real analysis, set theory and linear algebra which has been referred to in the text and has been generally assumed to be well known. We have included this material so that the notation, definitions and results referred to will be a little clearer for the student.

APPENDIX 1. THE REAL ANALYSIS BACKGROUND

A1.1 Notation. We denote by

$\mathbb{N}$ the set of natural numbers:

$$1,2,3,\ldots,n,\ldots$$

$\mathbb{Z}$ the set of integers:

$$\ldots,-n,\ldots,-2,-1,0,1,2,\ldots,n,\ldots$$

$\mathbb{Q}$ the set of rational numbers:

$$\frac{p}{q} \quad (p,q \text{ mutually prime } p \in \mathbb{Z} \text{ and } q \in \mathbb{N})$$

$\mathbb{R}$ the set of real numbers.

$\mathbb{Z}^+$, $\mathbb{Q}^+$ and $\mathbb{R}^+$ are the subsets of *positive* elements;   that is, those elements x such that $x \geqslant 0$.

The *strictly positive* elements are those elements x such that $x > 0$.

$\mathbb{Q}$ is an ordered field and is extended to the ordered field $\mathbb{R}$ which is defined by its possession of the following property.

A1.2 The Supremum Axiom. *Every non-empty subset which is bounded above has a least upper bound.*

The subsets of rational and irrational numbers embedded in the reals are each dense in the reals.

A1.3 The Density Properties. *Between any two real numbers there is*

(*i*)   *a rational number and*

(*ii*)   *an irrational number;*

that is, given x,y ∈ ℝ *where* x < y *there exists a rational number* r *and an irrational number* s *such that*

(*i*)   x < r < y *and*

(*ii*)   x < s < y.

The Supremum Axiom has an equivalent form expressed in terms of convergence of sequences.

A1.4 The Completeness of the Real Number System. *Every Cauchy sequence is convergent.*

So ℝ is called a *complete ordered field.*

The Supremum Axiom has a useful implication for the convergence of monotone sequences.

A1.5 The Monotone Convergence Theorem. *Every bounded monotone sequence is convergent.*

The development of real analysis depends basically on a compactness property.

A1.6 The Local Compactness of the Real Number System. *Every bounded sequence has a convergent subsequence.*

Of course the limit point of the convergent subsequence need not be a member of the original sequence. The property simply asserts the existence of the limit point in ℝ.

Bounded closed intervals are particularly significant because of this property.

A1.7 Corollary. *Every sequence in a bounded closed interval has a subsequence convergent to a point of the interval.*

The following properties of continuous functions are fundamental.

A1.8 The Connectedness Property. *A continuous function maps intervals to intervals.*

This is also called the *Intermediate Value Property*.

A1.9 The Compactness Preserving Property. *A continuous function on a bounded closed interval (i) is bounded and*

(ii) has a maximum and minimum value.

A1.8 and A1.9 have the following implication.

A1.10 Corollary. *A continuous function maps bounded closed intervals to bounded closed intervals.*

Continuous functions on bounded closed intervals obey a particularly strong continuity property.

A1.11 Heine's Theorem. *A continuous function on a bounded closed interval is uniformly continuous.*

This result implies that such functions can be approximated by a much simpler class of functions.

A1.12 Corollary. *For a continuous function f on [a,b], given $\varepsilon > 0$ there exists a step function g on [a,b] such that $|f(x)-g(x)| < \varepsilon$ for all $x \in [a,b]$.*

The following property is essential for developing the theory of differentiation.

A1.13 The Mean Value Theorem of differential calculus. *For a function f which is continuous on [a,b] and differentiable on (a,b) there exists a $c \in (a,b)$ such that*

$$f(b) - f(a) = f'(c)(b-a).$$

The student should be familiar with Taylor's Theorem and the two common forms of the remainder.

A1.14 Taylor's Theorem. *For a function* f *which is* n+1 *times differentiable on an open interval* J, *given* a,x $\in$ J

$$f(x) = f(a) + f'(a)(x-a) + \frac{f''(a)}{2!}(x-a)^2 + \ldots + \frac{f^{(n)}(a)}{n!}(x-a)^n + R_{n,a}(x)$$

where

$$R_{n,a}(x) = \frac{f^{(n+1)}(c_1)}{(n+1)!}(x-a)^{n+1} \quad \textit{for some } c_1 \textit{ between a and x,}$$

(the *Lagrange form* of the remainder)

or equivalently,

$$R_{n,a}(x) = \frac{f^{(n+1)}(c_2)}{n!}(x-c_2)^n(x-a) \quad \textit{for some } c_2 \textit{ between a and x,}$$

(the *Cauchy form* of the remainder).

We should be clear about our definition of the Riemann integral.

A1.15 Notation and Definition. For a bounded closed interval $[a,b]$, a finite subset $P \equiv \{x_0, x_1, \ldots, x_n\}$ of $[a,b]$ where $a = x_0 < x_1 < \ldots < x_n = b$ is called a *partition* of $[a,b]$.

Given a bounded function f on $[a,b]$ and writing

$$M_k(f) \equiv \sup\{f(x) : x \in [x_{k-1}, x_k]\}$$
$$m_k(f) \equiv \inf\{f(x) : x \in [x_{k-1}, x_k]\},$$

we define

$$U(P,f) \equiv \sum_{k=1}^{n} M_k(f)(x_k - x_{k-1})$$
$$L(P,f) \equiv \sum_{k=1}^{n} m_k(f)(x_k - x_{k-1})$$

the *upper and lower Riemann sums* for f with respect to partition P. We say that f is *Riemann integrable* if

$$\inf\{U(P,f) : \text{for all partitions } P\}$$

$$= \sup\{L(P,f) : \text{for all partitions } P\}$$

and the common value is denoted $\displaystyle\int_a^b f(t)dt$.

The following characterisation for integrability is useful in the development of the theory of integration.

A1.16 Riemann's Condition. *A bounded function f on* $[a,b]$ *is Riemann integrable if and only if given* $\varepsilon > 0$ *there exists a partition* P *of* $[a,b]$ *such that*

$$U(P,f) - L(P,f) < \varepsilon.$$

Corollary A1.12 defines a significant class of integrable functions.

A1.17 Theorem. *Every continuous function on a bounded closed interval is Riemann integrable.*

Central to the study of differential and integral calculus is the theorem which links the two different problem areas.

A1.18 The Fundamental Theorem of Calculus. *Consider an integrable function* f *on* $[a,b]$.

(i)   *If f is continuous at* $c \in (a,b)$ *then the function* F *on* $[a,b]$ *defined by*

$$F(x) = \int_a^x f(t)dt$$

is differentiable at c *and* $F'(c) = f(c)$.

(ii)   *If* F *is a function which is continuous on* $[a,b]$ *and differentiable on* (a,b) *and*

$$F'(x) = f(x) \quad \text{for all } x \in (a,b)$$

then $\displaystyle\int_a^b f(t)dt = F(b) - F(a)$.

APPENDIX 2.   THE SET THEORY BACKGROUND

A2.1 Definitions and Notation.

(i)   A basic relation between sets is *set inclusion*.
We say that A is a *subset* of B if every member of A is a member
of B and we write $A \subseteq B$.
We say that A is a *proper subset* of B if $A \subseteq B$ and there is a
member of B which is not a member of A and we write $A \subset B$.
Of course, A = B if $A \subseteq B$ and $B \subseteq A$;   this observation provides
us with the basic technique for proving set equality.

(ii)   When we consider a family of sets it is often convenient to
index the family. For a finite family we will use the convention $\{A_k : k \in \{1,2,\ldots,n\}\}$. For a family not necessarily
finite, we will use the notation $\{A_\alpha\}$ where α is a member of
some index set.

(iii)   Using index notation for a family of sets $\{A_\alpha\}$ we define the
union of sets in the family as

$$\cup A_\alpha \equiv \{x : x \in A_\alpha \quad \text{for at least one } \alpha\}$$

and the *intersection* as

$$\cap A_\alpha \equiv \{x : x \in A_\alpha \quad \text{for all } \alpha\}$$

(iv)   Given a specified universal set, the *complement* of a set A is
the set of elements which are not members of A and is denoted
by C(A).
Given sets A and B, the *set difference* of A less B is the set

$$A \setminus B \equiv \{x : x \in A \text{ and } x \notin B\}.$$

Complementation is related to the taking of unions and inter-
sections by the following basic result.

A2.2 de Morgan's Theorem. *For any non-empty family of sets* $\{A_\alpha\}$,

(*i*) $C(\cup A_\alpha) = \cap C(A_\alpha)$,

(*ii*) $C(\cap A_\alpha) = \cup C(A_\alpha)$.

A2.3 Definitions.

(i)   Given sets X and Y a *function* or *mapping* T of X into Y written

T : X → Y   where

y = Tx   or   x ⟼ Tx

assigns to each x ∈ X one and only one element Tx ∈ Y.

When the second set is the scalar field ℝ (or ℂ) we most often refer to these as functions and talk about a *real* (or *complex*) *function* f on X.

(ii)   For any subset A of X, the set

T(A) ≡ {y ∈ Y : y = Tx   for some x ∈ A}

is called the *image* of A under the mapping T.
We call X the *domain* and T(X) the *range* of the mapping T.
When T(X) = Y we say that T maps X *onto* Y.

(iii)   For any subset B of Y, the set

$T^{-1}(B) \equiv \{x \in X : y = Tx$ for some $y \in B\}$

is called the *inverse image* of G under the mapping T.
If T is not onto then for y ∉ T(X) we have $T^{-1}(\{y\}) = \phi$.
If y ∈ T(X) then $T^{-1}(\{y\})$ is not necessarily a single point set.
However, if for each y ∈ T(X), $T^{-1}(\{y\})$ is a single point set then we say that T is a *one-to-one* mapping.

Taking inverse images of sets under a mapping satisfies more regular properties than the taking of images. The basic properties are contained in the following result.

A2.4 Theorem. *Consider any mapping* T : X → Y.

(i)   *For any non-empty family* $\{A_\alpha\}$ *of subsets of* X,

(a) $T(\cup A_\alpha) = \cup T(A_\alpha)$,

(b) $T(\cap A_\alpha) \subseteq \cap T(A_\alpha)$.

(ii)   For any non-empty family $\{B_\beta\}$ of subsets of Y,

(a) $T^{-1}(\cup B_\beta) = \cup T^{-1}(B_\beta)$

(b) $T^{-1}(\cap B_\beta) = \cap T^{-1}(B_\beta)$.

A2.5 Definitions.

(i)   For a one-to-one mapping $T : X \to Y$ we can define an *inverse* mapping $T^{-1} : T(X) \to X$ which assigns to each $y \in T(X)$ the element $x \in X$ such that $y = Tx$.

(ii)   Given sets X, Y and Z and mappings $T : X \to Y$ and $S : Y \to Z$, the *composite* mapping $S \circ T : X \to Z$ assigns to each element $x \in X$ the element $S(Tx) \in Z$. It should be remembered that composition of mappings is associative but is not commutative.

(iii) The *identity* mapping on X, $\mathrm{id}_X : X \to X$ is defined by $\mathrm{id}_X(x) = x$.

A2.6 Theorem. *A mapping $T : X \to Y$ is one-to-one and onto if and only if there exists a mapping $S : Y \to X$ such that*

$$S \circ T = \mathrm{id}_X \quad and \quad T \circ S = \mathrm{id}_Y.$$

Then, of course, $S = T^{-1}$.

A2.7 Definition. Given a mapping $T : X \to Y$ it is sometimes convenient to consider T defined only on a proper subset A of X. In this case we refer to the *restriction* of T to A which is denoted by $T|_A : A \to X$ and is defined by $T|_A(x) = Tx$.
Given a proper subset A of X and a mapping $S : A \to Y$, any mapping $T : X \to Y$ where $T|_A = S$ is called an *extension* of S to X.

A2.8 Definitions.

(i)   Given sets X and Y the *product* (or *Cartesian product*) written $X \times Y$ is the set of all ordered pairs (x,y) where $x \in X$ and $y \in Y$. The order of the pairing indicates that $X \times Y$ and $Y \times X$ are different products when $X \neq Y$. But it is not necessary that X and Y be different sets: the product $\mathbb{R}^2 = \mathbb{R} \times \mathbb{R}$ is familiar as the Euclidean plane.

The concept is generalised to the product of n sets
$\{X_1, X_2, \ldots, X_n\}$.

$$\prod_{k=1}^{n} X_k \equiv X_1 \times X_2 \times \ldots \times X_n,$$

is the set of ordered n-tuples $(x_1, x_2, \ldots, x_n)$ where $x_k \in X_k$
for each $k \in \{1, 2, \ldots, n\}$.

We could generalise further to the product of a sequence of
sets $\{X_1, X_2, \ldots, X_n, \ldots\}$

$$\prod_{k \in \mathbb{N}} X_k \equiv X_1 \times X_2 \times \ldots \times X_n \times \ldots$$

The concept of the product of sets can be seen to be independent
of order if we consider the product of sets $\{X_\alpha\}$ as the set of
all possible mappings on the index set with the property that
the mapping takes each index α to an element $x_\alpha \in X_\alpha$. In this
way we can define the product ΠX_α.

(ii)   Given any product, say $\prod_{k=1}^{n} X_k$, for any n-tuple $x \equiv (x_1, x_2, \ldots, x_n)$
we call x_k the kth *co-ordinate* of x for each $k \in \{1, 2, \ldots, n\}$.
The natural mappings associated with any product are the
co-ordinate projection mappings. Given $k \in \{1, 2, \ldots, n\}$ the
mappings $p_k : \prod_{k=1}^{n} X_k \to X_k$ defined by

$$p_k(x) = x_k$$

is the kth *co-ordinate projection* mapping.

We are so accustomed to associating a function with its graph
that we tend to obscure the difference between the two objects.

A2.9 Definition. Given a mapping $T : X \to Y$ the *graph* of T, denoted by G_T is
the subset of $X \times Y$ defined by

$$G_T \equiv \{(x, y) : y = Tx\}.$$

Given a mapping $T : X \to Y$ it is possible to induce another mapping
$S : X \to X \times Y$ defined by

$$Sx = (x, Tx)$$

and S is always a one-to-one mapping of X onto G_T.

A2.10 Definitions.

(i)   Given a set X, a *relation* on X is a meaningful statement which, to each element $x \in X$ associates another element $y \in X$. (By meaningful we mean that for each $x \in X$ we can say whether or not any $y \in X$ is associated with x by the relation.)

(ii)   A basic type of relation is an *equivalence relation*. For such a relation we say that x is *equivalent* to y and we write $x \sim y$. An equivalence relation is defined by its properties:

(a) $x \sim x$ for every $x \in X$ (reflexivity)

(b) $x \sim y$ implies $y \sim x$ (symmetry)

(c) $x \sim y$, $y \sim z$ implies $x \sim z$ (transitivity).

(iii)   A *partition* of a set X is a family of non-empty subsets of X which are mutually disjoint and whose union is X.

There is a natural association between the equivalence relations on a set and partitions of the set.

A2.11 Theorem. *An equivalence relation on a set X generates a partition of X and a partition of X generates an equivalence relation on X.*

A2.12 Definition. The partition on X generated by an equivalence relation on X is called the family of equivalence classes on X. Given $x \in X$, the *equivalence class* [x] generated by x is the set

$$[x] \equiv \{y \in X : y \sim x\}.$$

The study of cardinality of sets is based on a special equivalence relation called numerical equivalence.

A2.13 Definitions.

(i)   Given the family of all subsets of a given universal set we
        say that a subset X is *numerically equivalent* to a subset Y if
        there exists a one-to-one mapping T of X onto Y.
        The equivalence classes of subsets which are numerically
        equivalent are called *cardinal numbers*.

(ii)   A subset X is said to be *countable* if it is numerically
        equivalent to a subset of the natural numbers $\mathbb{N}$.
        If there exists an n $\in$ $\mathbb{N}$ such that X is numerically equivalent
        to the set $\{1,2,\ldots,n\}$ then X is said to be *finite*.   A subset
        which is not finite is said to be *infinite*.
        A subset which is numerically equivalent to the set of natural
        numbers $\mathbb{N}$ is said to be *countably infinite*.
        A subset which is not countably infinite is said to be
        uncountable.

A2.14 Theorem. *Every infinite subset of a countably infinite set is*
countably infinite.

        The following property is important for developing the theory
of countably infinite sets.

A2.15 Theorem. *The product* $\mathbb{N} \times \mathbb{N}$ *is countably infinite.*

        This result has significant general consequences.

A2.16 Theorem. *The union of a countable family of countable sets is*
countable.

        The following cardinality properties for the real number system
$\mathbb{R}$ are basic.

A2.17 Theorem. *The set of rational numbers* $\mathbb{Q}$ *is countably infinite.*

A2.18 Cantor's Theorem. *The set of real numbers* $\mathbb{R}$ *is uncountable.*

A2.19 Corollary. *The set of irrational numbers is uncountable.*

APPENDIX 3.   THE LINEAR ALGEBRA BACKGROUND

A3.1 Definition.   A *linear space* X over $\mathbb{R}$ (or $\mathbb{C}$) is a non-empty set with operations *addition*:

for every $x,y \in X$, $x+y \in X$,   and

multiplication by a scalar:

for every scalar λ and $x \in X$, $\lambda x \in X$,

which satisfy the following properties:

(i) $x + (y+z) = (x+y) + z$ for all $x,y,z \in X$,

(ii) $x + y = y + x$ for all $x,y \in X$,

(iii)   there exists an element $\underset{\sim}{0} \in X$ such that

$$x + \underset{\sim}{0} = x,$$

(iv)   for each $x \in X$ there exists an element $-x \in X$ such that

$$x + (-x) = \underset{\sim}{0}$$

(v) $(\lambda+\mu)x = \lambda x + \mu x$ for all scalars λ,μ and $x \in X$,

(vi) $\lambda(x+y) = \lambda x + \lambda y$ for all scalars λ, and $x,y \in X$,

(vii) $\lambda(\mu x) = (\lambda\mu)x$ for all scalars λ,μ and $x \in X$,

(viii)   for scalar 1,

$$1x = x \quad \text{for all } x \in X.$$

A linear space is also known as a *vector space* and the elements of the linear space are called *vectors*.

We adopt the convention of writing the vectors in Latin alphabet and the scalars in Greek.   Whenever it appears that there may be confusion over notation we emphasise the vector elements by $\underset{\sim}{x}$:   for example we use $\underset{\sim}{0}$ to distinguish the zero vector   from 0 the zero scalar.

A linear space over $\mathbb{R}$ is often called a *real linear space* and a linear space over $\mathbb{C}$ is often called a *complex linear space*.   A complex linear space is also a real linear space if we restrict multiplication by a scalar to real scalars.

A subset A of a linear space X is called a *linear subspace* of X
if (i) $x + y \in A$ for all $x, y \in A$, and

(ii) $\lambda x \in A$ for all scalars λ and $x \in X$.

A3.2 Definitions. Given finite subsets $\{x_1, x_2, \ldots, x_n\}$ in X and scalars
$\{\lambda_1, \lambda_2, \ldots, \lambda_n\}$, the vector $\sum_{k=1}^{n} \lambda_k x_k$ is called a *linear combination* of
$\{x_1, x_2, \ldots, x_n\}$.

Given any subset A of X the *span* of A denoted sp A, is the set of all
linear combinations of elements from A.

A subset A of X is said to be *linearly independent* if for every
finite subset $\{x_1, x_2, \ldots, x_n\}$ in A and scalars $\{\lambda_1, \lambda_2, \ldots, \lambda_n\}$, we have that
$\sum_{k=1}^{n} \lambda_k x_k = \underline{0}$ implies that $\lambda_1 = \lambda_2 = \ldots = \lambda_n = 0$.
A linearly independent subset A of X such that sp A = X is called a *basis*
for X, (a *Hamel* basis).

The notion of basis is crucial in the development of linear
algebra because of the following uniqueness property.

A3.3 Theorem. *A non-empty subset* A *of a linear space* X *is a basis for* X
if and only if every element of X *can be expressed uniquely as a linear*
combination of elements of A.

A3.4 Definition. A linear space X is said to be *finite dimensional* if
X = $\{\underline{0}\}$ or X has a finite basis. Otherwise X is said to be *infinite*
dimensional. In general, we are not interested in the trivial linear
space X = $\{\underline{0}\}$.

Any discussion of finite dimensional linear spaces depends
essentially on the following cardinality property.

A3.5 Theorem. *For a finite dimensional linear space* X *all bases have the*
same number of elements.

A3.6 Definition. For a finite dimensional linear space X the number of
elements in a basis is called the *dimension* of X.

The natural mappings between linear spaces are the linear
mappings, often called in algebra texts the linear space *homomorphisms*.

A3.7 Definition. Given linear spaces X and Y over the same scalar field, a mapping $T : X \to Y$ is called a *linear* mapping if

  (i) $T(x_1 + x_2) = Tx_1 + Tx_2$ for all $x_1, x_2 \in X$
                    (the additive property),

and

  (ii) $T(\lambda x) = \lambda Tx$ for all $x \in X$ and scalars λ
                    (the homogeneity property).

In particular, a linear mapping $T : X \to X$ is often called a linear *operator* on X and a linear mapping f of X into its scalar field is called a linear *functional* on X. We generally use a lower case letter for mappings into the scalar field.

Given linear spaces X and Y over the same scalar field and a linear mapping $T : X \to Y$, it is clear that for any linear subspace A of X, the image T(A) is a linear subspace of Y and for any linear subspace B of Y, the inverse image $T^{-1}(B)$ is a linear subspace of X. In particular, $T(\underline{0}) = \underline{0}$ and the *kernel* of T, denoted by ker T, defined by ker $T = T^{-1}(\{\underline{0}\})$, is a linear subspace of X.

An important property of linear mappings is given through the kernel.

A3.8 Theorem. *Given linear spaces X and Y over the same scalar field a linear mapping $T : X \to Y$ is one-to-one if and only if ker $T = \{\underline{0}\}$.*

A3.9 Definition. Given linear spaces X and Y over the same scalar field, a linear mapping $T : X \to Y$ is called an *isomorphism* if T is one-to-one. The linear spaces X and Y are said to be *isomorphic* if there exists an isomorphism T of X onto Y.

It is important to see how the structure of n-dimensional linear spaces is that of $\mathbb{R}^n$ (or $\mathbb{C}^n$).

A3.10 Theorem. *An n-dimensional linear space X_n over $\mathbb{R}$ (or $\mathbb{C}$) is isomorphic to $\mathbb{R}^n$ (or $\mathbb{C}^n$).*

A3.11 Remarks. Consider basis $\{e_1, e_2, \ldots, e_n\}$ for X_n For any $x \in X_n$, $x = \lambda_1 e_1 + \lambda_2 e_2 + \ldots + \lambda_n e_n$ (uniquely by Theorem A3.3), and an isomorphism of X_n onto $\mathbb{R}^n$ (for $\mathbb{C}^n$) is given by

$$x \longmapsto (\lambda_1, \lambda_2, \ldots, \lambda_n).$$

A3.12 Definition. Given a family of linear spaces $\{X_\alpha\}$ over the same scalar field, the product space πX_α has a natural linear structure where the linear operations are defined co-ordinatewise; that is, we define addition by:

$$x_\alpha + y_\alpha = (x+y)_\alpha \quad \text{for all } x,y \in X_\alpha,$$

and multiplication by a scalar by:

$$\lambda x_\alpha = (\lambda x)_\alpha \quad \text{for all scalars } \lambda \text{ and } x \in X_\alpha.$$

A3.13 Definition. Given a linear space X and a proper linear subspace M, an equivalence relation which occurs naturally on X is that defined by

$$x \sim y \quad \text{if} \quad x - y \in M.$$

The equivalence classes are called *cosets* and

$$[x] = x + M \quad \text{for every } x \in M.$$

Now the family of cosets, denoted $\frac{X}{M}$ inherits a linear structure from X. We define

$$[x] + [y] = [x+y] \quad \text{for all } x,y \in X$$

and multiplication by a scalar by:

$$\lambda[x] = [\lambda x] \quad \text{for all scalars } \lambda \text{ and } x \in X$$

We call the linear space $\frac{X}{M}$ a *quotient* (or *factor*) space.

Given linear spaces X and Y over the same scalar field and linear mapping $T : X \to Y$, $\frac{X}{\ker T}$ is a commonly occurring quotient space. The mapping T induces a linear mapping $\tilde{T} : \frac{X}{\ker T} \to Y$ defined by

$$\tilde{T}[x] = Tx$$

and $\widetilde{T}$ is one-to-one on $\frac{X}{M}$, (see Theorem A3.8).

A3.14 Definition. A linear space X over $\mathbb{R}$ (or $\mathbb{C}$) is said to be an *algebra* if there exists an operation *multiplication*

$$\text{for every } x,y \in X, \quad xy \in X$$

which satisfies the following properties:

  (i) $x(yz) = (xy)z$ for all $x,y,z \in X$,

  (ii) $(\lambda x)(\mu y) = (\lambda\mu)(xy)$ for all scalars λ,μ and $x,y \in X$,

  (iii) $x(y+z) = xy + xz$ and
           $(y+z)x = yx + zx$ for all $x,y,z \in X$.

X is said to be a *commutative* algebra if also

  (iv) $xy = yx$ for all $x,y \in X$.

An element $e \in X$ is said to be an *identity* if

  (v) $ex = xe = x$ for all $x \in X$.

          A subset A of an algebra X is said to be a *subalgebra* of X if A is a linear subspace of X and $xy \in A$ for all $x,y \in A$.

INDEX TO NOTATION

INDEX